COMBINABLE CROPS AND THE EC

A Guide to European Community
Price Support Mechanisms
and Legislation Affecting the Marketing of Cereals,
and Other Combinable Crops

Dr Mary Abbott

BSP PROFESSIONAL BOOKS

OXFORD LONDON EDINBURGH

BOSTON MELBOURNE

Copyright © National Farmers Union 1990

First published 1990

British Library
Cataloguing in Publication Data
Abbott, Mary
   Combinable crops and the EC.
   1. European Community countries.
   Agricultural products. Prices. Policies of
   European Economic Community
   I. Title
   338.1'81

   ISBN 0–632–02814–9

BSP Professional Books
A division of Blackwell Scientific
   Publications Ltd
Editorial Offices:
Osney Mead, Oxford OX2 0EL
   (Orders: Tel. 0865 240201)
25 John Street, London WC1N 2 BL
23 Ainslie Place, Edinburgh EH3 6AJ
3 Cambridge Center, Suite 208, Cambridge
   MA 02142, USA
107 Barry Street, Carlton, Victoria 3053,
   Australia

Set by DP Photosetting, Aylesbury, Bucks
Printed and bound in Great Britain by
Hollen Street Press Ltd, Slough

Contents

Foreword

by Sir Simon Gourlay, President, National Farmers Union

The NFU Handbook, *Combinable Crops and the EC*, is the first volume of a new series of titles published by Blackwell Scientific Publications (BSP Professional Books).

This book is aimed primarily at producers of cereals and other combinable crops to act as a handy reference volume explaining how the EC works, the complexities of Community price support for cereals, oil seeds and other combinable crops, and the debate on the future of the CAP.

The book should also prove useful to traders, processors, advisors and students – in fact, helpful to anyone with a vested interest in understanding how combinable crops are affected by the European Community.

The NFU deals with hundreds of enquiries a year from members asking questions on the EC, the CAP and the price support arrangements. It is hoped that this volume will answer many of these queries.

The author, Dr Mary Abbott, is one of the NFU's most experienced specialists, and a past Director of the NFU's Arable Division.

Sir Simon Gourlay
President, NFU

1 The European Community Decides

The transfer of farm price support to the European Economic Community has created the situation where the mechanisms of support, the fixing of prices, and the overall management of farm product markets has become somewhat perplexing to most observers.

The UK joined the EC in 1973 and the transitional arrangements were fully completed by the beginning of 1978 (see Table 1.1). Unfortunately, as each year has passed, the EC has contrived to make matters more complicated, adding layer upon layer of new arrangements. In addition, the reasons for the Community's decisions may not be apparent, and may even appear perverse. Not only does this cause great confusion for individuals, but it means that farmers are not being given clear signals for the future direction of farm support policy.

It may be that the nature of the European Community makes such problems almost inevitable. The objectives defined in the Treaties are in places contradictory, and open to different interpretation; and with 12 member states the various national interests have to be reconciled. As a result, complexity is added as extra elements are put in to satisfy one member state or another, and decisions taken by the Community are usually compromises. This has an important effect on farm policy.

Table 1.1 The European Community.

Key Developments
  1957 – European Economic Community established by the Treaty of Rome.
    Members: France, Germany, Italy, Netherlands, Belgium, Luxembourg.
  1973 – UK, Ireland and Denmark join.
  1978 – Transitional arrangements completed 1 January: UK producers
    now totally dependent on EC arrangements.
  1979 – First direct elections to the European Parliament.
  1981 – Greece joins.
  1986 – Spain and Portugal join.
  1987 – Single European Act amends earlier Treaties.

1.1 Setting up the European Economic Community: The Treaty of Rome

The treaty setting up the European Economic Community was signed in Rome on 25 March 1957. The original signatories were France, Germany, Italy, Netherlands, Belgium and Luxembourg.

The European Economic Community, with the earlier established European Coal and Steel Community, and the European Atomic Energy Community, together constitute the European Communities (EC).

The Treaty of Rome is a document which is wide ranging. Its stated aim is to promote throughout the Community a harmonious development of economic activities, a continuous and balanced expansion, and closer relations between the member states. In practical terms, the broader objectives have been largely eclipsed until recently by the Common Agricultural Policy and the difficulties it has faced.

The Treaty of Rome provided the basis for setting up the Common Agricultural Policy and, in Article 39, defined its objectives:

(a) to increase agricultural productivity by promoting technical progress and by ensuring the rational development of agricultural production and the optimum utilisation of the factors of production, in particular labour;
(b) thus to ensure a fair standard of living for the agricultural community, in particular by increasing the individual earnings of the persons engaged in agriculture;
(c) to stabilise markets;
(d) to assure the availability of supplies;
(e) to ensure that supplies reach consumers at reasonable prices.

As new member states joined the EC (UK, Denmark and Ireland in 1973, Greece in 1981, Spain and Portugal in 1986) Accession Treaties defined the arrangements for transition to full membership.

1.2 The Single European Act and 1992

The Single European Act of 1987 made major amendments to the Treaty of Rome. In signing this treaty, the member states all undertook significant new commitments, the most important being the creation of the Single European Market by the end of 1992. This was defined in the following terms:

'The internal market shall comprise an area without frontiers in which free movement of goods, persons, services and capital is ensured in accordance with the provisions of this Treaty.'

In addition, the Single European Act contains a range of new objectives for the EC. These include:

- convergence of economic and monetary policies;
- improvement of health and safety at work;
- reducing regional disparities, supported by EC financing;
- strengthening of the scientific and technological base of European industry;
- environmental action with the objectives of preserving, protecting and improving the quality of the environment, contributing to protecting human health, and ensuring a prudent and rational use of natural resources.

The EC's decision-making procedures were revised to try to ensure that the necessary adaptation of its legislation could be agreed for 1992. Majority voting became the normal method of deciding legislation and the methods of delegating powers to the Commission were streamlined. The European Councils ('Summits') were made a formal Community arrangement.

The Commission's idea of a single market includes the abolition of all internal frontier customs posts and checks and the harmonisation of indirect taxation. In terms of technical standards, the EC will define minimum standards, and member states will recognise one another's regulations as equivalent to their own. Freedom of trade is supported by the Community's competition policy.

Trade in agricultural products within the EC is already relatively free of restrictions. Work has been under way for many years to reduce barriers caused by considerations of public health, labelling or other restrictions. Nevertheless, the move towards a single market has a number of potential implications for the arable sector.

One of the areas directly affected is that of plant health. In order to remove trade barriers, and yet maintain plant health standards, the Commission is aiming to introduce common standards for plant health, with a Community Plant Health Inspectorate, and certification at the point of production to allow free movement throughout the EC. Pests and diseases would be defined as confined to certain zones, not to individual countries, and necessary emergency action would be taken by the Commission, rather than by national authorities.

The Single European Act also has important financial consequences. The commitments of the Single European Market contributed to the need to develop a new basis for financing the EC (see Chapter 8). The Single Market should lead to national aids being better controlled, but there is now increased emphasis on income aids for smaller farmers and the use of EC funds to support rural development in regions most affected by CAP reform. Limits are now applied to the share of funds taken by agriculture, and monetary compensatory amounts MCAs are due to be abolished as the green money system is brought into line with general money arrangements (see Chapter 8).

1.3 The Council

The Council (the Council of Ministers) is the principal decision-taking body of the European Community. Its responsibilities are defined by the Treaty of Rome. One Minister from each member state sits on the Council.

The Council is chaired by the President. The office of President of the Council is held by each member state in turn for six months. The sequence of the Presidency has been determined alphabetically, according to the names of the countries in their own language: 1987: Belgium, Denmark; 1988: Germany, Greece; 1989; Spain, France; 1990: Ireland, Italy; 1991: Luxembourg, Netherlands; 1992: Portugal, UK. During the following six years the pairs in each year are reversed so that the order becomes: Denmark, Belgium, Greece, Germany, France, Spain, Italy, Ireland, Netherlands, Luxembourg, UK, Portugal.

Although in relation to the CAP the term 'Council' usually means the meetings of Agriculture Ministers, it equally may refer to meetings of Foreign, Finance, Interior or other Ministers. The European Council is the 'Summit' meeting of Heads of State or of Government, which takes place at least twice a year.

The work of the Council, and its meetings, are prepared by the Committee of Permanent Representatives (COREPER) which comprises the Member States' ambassadors to the Community, each assisted by national civil servants. Preparation for Agriculture Council discussions on structures, agri-monetary or commodity issues is delegated to the Special Committee on Agriculture. A Council Secretariat carries out the necessary administrative work for the activities of the Council.

The Council is responsible for all major EC legislation, but delegates to the Commission much of the detailed implementation. Thus, the legislation for a particular scheme is likely to be contained in more than one set of regulations:

- A Council regulation will give the broad outlines – for instance, that grain shall be purchased into intervention and the price at which that will take place, but specifying only a limited number of other criteria for intervention purchases.
- A number of Commission regulations define the details by which the Council's regulation will be implemented.

The Council's voting procedures are established in the Treaty. Only the Council members may vote, not the Commission. As a consequence of the Single European Act, the normal procedure is now for decisions to be taken by majority voting. The opportunities for a single country to veto a decision have been largely curtailed, though some exceptions remain. Where decisions are taken by qualified majority, the votes of different member state

Table 1.2 Qualified majority voting.

Votes are weighted as follows:	
Belgium	5
Denmark	3
Germany	10
Greece	5
Spain	8
France	10
Ireland	3
Italy	10
Luxembourg	2
Netherlands	5
Portugal	5
UK	10
Total	76

representatives are weighted as shown in Table 1.2.

The Treaty lays down the procedures by which the Council will come to decisions on different matters. So far as commodity policies are concerned, the Council normally acts upon a proposal from the Commission. Then, to be agreed, this must be supported by at least 54 votes. In other cases there may have to be at least 54 votes cast by at least eight members. Unanimity is required if the Council amends a Commission proposal.

Major proposals before the Council are usually referred to the European Parliament, although only an opinion is required in many cases. However, the Council may also, in certain other cases, have to consult the European Parliament under the cooperation procedure introduced by the Single European Act. Under this arrangement, if the European Parliament rejects the Council's decision, the Council can only proceed if it has the unanimous support of all its members. If the Parliament proposes amendments, these will be considered by the Commission and the Council. In order for these amendments to be incorporated into Community law, those adopted by the Commission require qualified majority support in the Council; those rejected by the Commission require the unanimous support of the Council.

1.4 The Commission

The Commission has a number of roles: it makes proposals for future legislation for consideration by the Council, it takes decisions and implements legislation under powers which are delegated to it by the Council, and it generally has the task of ensuring that the legislation of the European Community is properly applied.

The 17 members of the Commission (listed in Table 1.3) are appointed for a four-year term. They include two nationals from each of the larger member states (Germany, France, Italy, Spain and UK) and one each from the other

Table 1.3 The Commission.

Member	Country	Special responsibilities
Jacques Delors	France	President Monetary affairs
Frans Andriessen	Netherlands	Vice President External relations and trade
Henning Christophersen	Denmark	Vice President Economic and financial affairs
Manuel Marin	Spain	Vice President Cooperation and development, Fisheries
Filippo Maria Pandolfi	Italy	Vice President Research, Science IT
Martin Bangemann	Germany	Vice President Internal market
Sir Leon Brittan	UK	Vice President Competition policy Financial institutions
Carlo Ripa Di Meana	Italy	Environment Nuclear safety
Antonio Cardoso E. Cunha	Portugal	Administration Energy, Euratom, etc.
Abel Matutes	Spain	Mediterranean policy
Peter Schmidhuber	Germany	Budget, financial control
Christiane Scrivener	France	Taxation
Bruce Millan	UK	Regional policy
Jean Dondelinger	Luxembourg	Communications
Ray MacSharry	Ireland	Agriculture Rural development
Karel van Miert	Belgium	Transport, credit Consumer protection
Vasso Papandreou	Greece	Social affairs, employment, etc.

member states. The President and six Vice-Presidents are appointed for a term of two years.

Commission decisions are taken by voting on the basis of one man, one vote, and individual Commissioners are required to be independent of their own governments.

The secretariat of the Commission, the Commission Services, is organised into 23 'directorates general' (DGs), headed by the Secretary General. DG VI deals with agriculture, but a number of others may also be relevant – for example, DG I (External Relations), DG III (Internal Market and Industrial Affairs), DG XI (Environment, Consumer Protection and Nuclear Safety), DG XII (Science, Research and Development) and DG XIX (Budgets). The senior staff structure of the Commission and DG VI is shown in Table 1.4.

The vast bulk of Community legislation concerning the CAP is written in Commission regulations. These describe not only the details of how support schemes operate, but also the week by week decisions on rates of aid, export refunds, etc. Consideration of these decisions involves an immense amount of work.

The Commission is assisted in its administration of commodity policies by Management Committees. These have been set up for each major commodity regime; in addition joint Management Committees deal with monetary compensatory amounts and trade mechanisms.

Table 1.4 Senior staff of the Commission (including those with specific responsibility for arable crops).

Commission	Secretary General: David Williamson
Agriculture (DGVI)	Director General: Guy Legras
Commodity policy	Deputy Director General: vacancy
Crop products	Directorate C: Organisation of markets in crop products
	Head: T.L.W. Windle Chief Advisor: I. de Gruben
	(1) Cereals, processed products and rice Head: R. Reifenrath Deputy: M. Thibault
	(2) Animal food and cereal substitutes Head: J. Sousa Uva
	(3) Sugar Head: E. Stendevad
	(4) Olive oil, olives and fibre plants Head: F. Gencarelli
	(5) Oilseeds and protein plants Head: Russell Mildon

Each Management Committee operates under powers defined by Council Regulations. The members of the Committee are civil servants from each member state and the meetings are chaired by a Commission official who does not have voting power.

Measures which the Commission intends to operate are submitted in draft to the appropriate Management Committee. The voting of Management Committees is weighted in the same way as the voting in the Council (see Table 1.2). A majority requires 54 votes.

Only a 'negative vote' of 54 or more votes against prevents the Commission from proceeding with a proposal. If the Commission wishes to go ahead with a proposal which has received a negative vote in the Management Committee, it must refer it to the Council, which has a month to decide (by qualified majority) whether or not to over-rule the Commission's intentions.

The Commission has set up a number of Advisory Committees which it consults on problems concerning the various market organisations. The members are appointed by the Commission on the basis of nominations by Community organisations representing agricultural producers, agricultural co-operatives, the agricultural and food industries, traders, workers and consumers. NFU representatives are nominated by COPA.

1.5 The European Parliament

Members of the European Parliament (MEPs) are elected for terms of five years. The numbers elected in each member state is given in Table 1.5. The members themselves elect the Parliament's President.

The Parliament normally takes decisions by simple majority voting, but it

Table 1.5 Numbers of MEPs elected by each member state.

Belgium	24
Denmark	16
Germany	81
Greece	24
Spain	60
France	81
Ireland	15
Italy	81
Luxembourg	6
Netherlands	25
Portugal	24
UK	81
Total	518

Table 1.6 UK members of the European Parliament Agriculture Committee.

Paul F. Howell	Conservative	Norfolk
Henry McCubbins	Labour	Scotland North East
Stan Newens	Labour	London Central
The Lord Plumb	Conservative	Cotswolds
Thomas Spencer	Conservative	Surrey West
George W. Stevenson	Labour	Staffordshire East

In addition there are the following substitutes:
Conservatives: Lord Inglewood, James Scott-Hopkins, Richard Simmonds
Labour: John Hume, David Morris, Barry Seal, Anthony Wilson.
Official Unionist Party (Christian Democratic Group): James Nicholson

Labour members, and Conservatives, belong respectively to the Socialist Group, and the European Democratic Group in the European Parliament.

does not have legislative powers in the same way as a national parliament since most important EC law is made by the Council. Its powers have, however, been increased as a consequence of the Single European Act, and, as explained earlier, in many cases – particularly concerning the establishment of the Single Market – the Parliament may have to give its opinion before regulations can be adopted.

But the Parliament has some important general powers: it can, with a two-thirds majority, force the Commission to resign, it may also reject the draft budget, and ask for a new draft to be submitted. The Parliament also has to agree to new member states being admitted to the EC. In addition, individual MEPs may table questions to the Commission in order to highlight issues, or to elicit useful information.

The Parliament meets in Strasbourg, for one week in each month, except August, and has an extra session in October to consider the budget. It also has a number of committees (meeting in Brussels), which carry out detailed work, such as examining specific proposals or issues. Table 1.6 lists the UK members of the Agriculture Committee.

1.6 The Economic and Social Committee

The Economic and Social Committee (ECOSOC) is an advisory body, established by the Treaty of Rome, which the Council and Commission have to consult on certain topics. It consists of the representatives of employers, workers and various interest groups such as agriculture, transport, trade, small businesses, the professions and consumers. There are 189 members, including 24 from the UK, appointed by the Council on the basis of nominations from member states. The Committee has a specialist section dealing with agriculture. Table 1.7 lists the UK members of that section.

Table 1.7 UK members of the Economic and Social Committee Agriculture Section.

Mr Jack R Body (General Secretary, National Union of Agricultural and
  Allied Workers)
Mr Kenneth J Gardiner (Consultant in food, law, technology and food
  technology)
Mr Colin A Hancock (Management Consultant)
Dr Peter Storie-Pugh (Former President of the Royal College of Veterinary
  Surgeons)
Mr Michael P Strauss (Co-ordinating Director, Policy, National Farmers' Union,
  retired)

1.7 The Court of Justice

The Court of Justice comprises 13 judges, appointed by consent of the
member state governments, for a period of six years. They are assisted by six
Advocates General, whose opinions in cases brought before the Court play
an important part in its eventual judgements.

It has the duties of interpreting and ensuring the proper application of the
Treaty of Rome by the individual member states. It also reviews the legality
of Council and Commission actions. If questions arise in national courts
about the proper functioning of the European Community, these may be
referred to the Court of Justice for a decision.

If the Commission considers that a member state has failed to fulfil a
Treaty obligation it will deliver a 'reasoned opinion' on the matter, after
giving the member state the opportunity to submit its observations. If the
member state does not comply with the reasoned opinion, within the
specified timetable, the Commission may bring the matter before the Court
of Justice. Member states may also bring actions before the Court of Justice
if the Commission fails to do so.

Parties to a case must comply with the Court's judgements.

1.8 The Court of Auditors

The Court of Auditors comprises 12 members appointed by the Council for
terms of six years.

It has the responsibility of auditing all EC revenue and expenditure. This
means examining the financial management of EC funds and expenditure
incurred under the various EC schemes. It involves its staff in making
detailed investigations within individual member states. Failure to operate
strictly within the terms of an EC regulation may result in the member state
being denied reimbursement for the expenditure it has incurred. Deliberate
fraud may be dealt with through national courts, or by the Court of Justice.

1.9 The farm price fixing: the decision-making process

Each year the Council takes decisions on farm support prices and related matters. The process of reaching those decisions is a major exercise, involving not only the Commission and Council, and other Community institutions that have to be consulted, but also those representing the interests of people whose lives, and businesses, can be significantly affected by the decisions that are taken.

In the autumn of each year, organisations such as COPA take a view of the next year's prices and make their demands known to Commission officials who are starting to draft the proposals. These drafts will be considered by the Commission towards the end of the year. Their content may well be leaked to the press, but they are not yet formal Commission proposals. It is usually January by the time the Commission has reached its decisions, and the 'Commission's proposals' are published.

The proposals come as three volumes. Volume I, an 'explanatory memorandum' of about 150 pages, reviews the previous year's developments and explains the proposals. Volume II details their financial implications. Volume III (200–300 pages usually) contains the draft regulations which, if approved by the Council, would enact the proposals.

The Council considers the Commission's proposals in a series of meetings from January onwards. The intention is that this process should be completed before the commencement of the new livestock product marketing years on 1 April. However, in recent years, the Council has frequently failed to meet this deadline. As a consequence, it has had to enact provisional legislation to deal with the situation, usually by rolling forward the previous year's prices for a limited period.

It has also failed to meet the deadlines for the arable crop products (particularly as southern member states' marketing years start on 1 June). In 1985, when the Council failed to agree any prices for cereals and rapeseed, the Commission imposed prices, in view of its responsibilities under the Treaty of Rome.

In parallel with the Council's deliberations, the European Parliament will consider the proposals. The Council has an obligation to consult the Parliament.

The Council will aim to find a package of measures which is acceptable to all member states. To do this, the President of the Council will hold a series of consultations with individual member states ('bilaterals'). During the course of discussion, a number of compromise packages may be considered: these may be presented by either the Commission or the Presidency, though the two will normally consult each other before anything is tabled.

In the Council meetings, Ministers may be assisted by their junior Ministers, Permanent Secretaries, and other senior advisers. Discussions are frequently prolonged, lasting over several days and nights, particularly in the final stages. The President may decide to restrict the Council session to Ministers only, excluding all advisers if it is felt that this will open the way to finding a compromise agreement. Throughout the price discussions, Ministers will be briefing the press and others.

When the price package is agreed, the legal texts have to be prepared. Where the outcome differs from the Commission's proposals, and if the changes have not already been spelt out in detail on paper, different interpretations may emerge of the agreement reached. Formal ratification may be carried out by Council meetings other than those of agriculture ministers. The published legal texts provide final confirmation, but there may be some delay before publication as the texts in the different languages have to be reconciled.

The Council regulations may have to be supplemented by Commission regulations agreeing the detail of any new scheme. These will be discussed in special working groups. Only when all these procedures have been completed can there be absolute certainty about the detailed content of the prices package.

Regulations normally become effective after publication in the Official Journal.

1.10 EC legislation applied to the UK

As a member of the European Community, the UK is subject to EC legislation. Two main categories of EC legislation are operated – regulations and directives:

- A *regulation* is a law which is directly applicable, in its entirety, in all member states.
- A *directive* sets out the results which must be achieved, leaving to the individual member states the choice of the method of doing this.

The European Communities Act (1972) has the effect of incorporating all EC regulations into UK law. However, new UK legislation may be necessary to implement EC directives in this country, and may also be necessary to bring into operation certain aspects of regulations, for instance where member states are required to take steps to ensure that schemes are properly operated, perhaps requiring penalties, or specific records to be kept.

When legislation is required, Parliament has the opportunity to debate it, but rejection of necessary legislation by the House, and failure to implement EC legislation properly, would be likely to place the UK in breach of its obligations under the Treaty of Rome, unless a *derogation* was granted.

Derogations may be allowed, exempting a member state from particular regulations, or parts of a regulation, or applying different conditions. Derogations have to be justified by the member state seeking them, are not

Table 1.8 House of Commons Select Committee on European Legislation.

Nigel Spearing (Chairman)	Newham South	Labour
George J. Buckley	Hemsworth	Labour
William Cash	Stafford	Conservative
Hugh Dykes	Harrow East	Conservative
Alan Haselhurst	Saffron Waldon	Conservative
Robert Hicks	Cornwall SE	Conservative
Jimmy Hood	Clydesdale	Labour
Michael Knowles	Nottingham E	Conservative
David Knox	Moorlands	Conservative
David Madel	Bedfordshire SW	Conservative
Tony Marlow	Northampton N	Conservative
Alan Meale	Mansfield	Labour
Bowen Wells	Hertford and Stortford	Conservative
Jimmy Wray	Provan	Labour

granted freely, and are usually granted only for a fixed time period.

The House of Commons Select Committee on European Legislation monitors all Community business considered by the Council. If it takes the view that issues of political importance arise, it will make recommendations for Commission proposals and other documents to be debated in the House of Commons. Whilst, conventionally, Ministers are not supposed to take decisions in Brussels on proposals that are awaiting debate in Westminster, such an approach is often impractical, especially since the passing of the Single European Act permits many decisions to be taken by majority voting. The membership of the Select Committee is given in Table 1.8.

The House of Commons Agriculture Committee, which gives detailed attention to particular chosen agricultural topics, will also review certain EC agricultural matters. This Committee will usually invite evidence from MAFF and various interested parties, including the NFU, and produce a

Table 1.9 House of Commons Select Committee on Agriculture.

Jerry Wiggin (Chairman)	Weston-super-Mare	Conservative
Richard Alexander	Newark	Conservative
Alan Amos	Hexham	Conservative
Christopher Gill	Ludlow	Conservative
Martyn Jones	Clwyd SW	Labour
Calum A. MacDonald	Western Isles	Labour
Seamus Mallon	Newry & Armagh	SDLP
Paul Marland	Gloucestershire W	Conservative
Eric Martlew	Carlisle	Labour
Elliot Morley	Glanford & Scunthorpe	Labour
Mrs Ann Winterton	Congleton	Conservative

Table 1.10 House of Lords European Communities Select Committee – Sub-Committee D (Agriculture and Food).

Lord Middleton (Chairman)
Lord Carter
Viscount Brookeborough
Baroness Elliot of Harwood
Baroness Gardner of Parkes
Lord Mackie of Benshie
Lord Northbourne
Earl Radnor
Lord Raglan
Duke of Somerset
Lord Stodart of Leaston

lengthy report and recommendations. The membership of the Agriculture Committee is given in Table 1.9.

The House of Lords European Communities Select Committee deals with agricultural business, including EC proposals, in its Sub-Committee D (Agriculture and Food). On chosen topics it will take written and oral evidence and report in full. The membership is given in Table 1.10.

1.11 EC lobby organisations

The NFU, in conjunction with the NFU of Scotland and Ulster Farmers' Union, maintains an office in Brussels – the British Agricultural Bureau (BAB) – with an executive staff of three. This provides a pivot for the UK farm lobby in Brussels; a substantial amount of representational work is carried out by UK-based people visiting Brussels, including farmers union office holders, committee chairmen and staff.

The principal organisation representing EC farmers is COPA. This, together with its sister organisation COGECA, representing cooperatives, is based in Brussels with a joint secretariat of over 40 staff. It is financed by its member organisations, which from the UK are the NFU, NFU of Scotland, and Ulster Farmers' Union. Policy issues are discussed at frequent committee sessions, with separate meetings for cereals, oilseeds, peas and beans, seeds, biotechnology etc. COPA representatives attend the Commission's advisory meetings. The Praesidium, comprising the presidents of the various member organisations, regularly meets the Agriculture Commissioner. Table 1.11 gives details of UK representation on COPA/COGECA.

Table 1.12 lists other EC lobby organisations, with a special interest in combinable crops, and to which UK trade organisations belong, together with the current representatives.

Table 1.11 COPA/COGECA: List of relevant committees and chairmen.

Committee	Chairman
Cereals	Mr J.J. Vorimore (French farmer, AGPB)
Oilseeds	Mr Robert Soucat (French farmer, AGPO)
Peas and Beans	Mr Marcel Carrière (French cooperative director)
Seeds	Dr W. Sohn (German Farmers' Union, DBV)
Biotechnology	Peter Shearer (UK; lately NFU Sugar Beet Secretary)

The UK is represented on these committees by delegates from the NFU, NFU of Scotland and Ulster Farmers' Union (usually, in each case, as appropriate the Chairman/Convenor and/or Secretary of the relevant Committee). The current Chairmen are:

NFU:
Cereals Committee:	A.W.D. (Tony) Pexton
Oil Protein and Fibre Crops Committee:	Jeremy Dillon-Robinson
Seeds Committee:	John H.E. Wells

NFU Scotland:
Cereals Committee:	G.A. (Sandy) Mole

UFU:
Cereals Committee:	Hugh Linehan

Also, COGECA representatives from FAC.

In addition, arable issues will be included in the agendas of the Praesidium (on which the UK is represented by the Presidents of the three farmers' unions), and the General Experts (on which the UK is represented by the permanent staff of BAB).

Abbreviations:
AGPB	Association Générale des Producteurs de Blé et autres Céréales
AGPO	Association Générale des Producteurs d'oléagineux
COGECA	Comité Général de la Cooperation de la CEE
COPA	Comité des Organisations Professionelles Agricoles de la CEE
DBV	Deutscher Bauernverband e V
FAC	Federation of Agricultural Cooperatives Ltd
NFUS	National Farmers' Union of Scotland
UFU	Ulster Farmers' Union

Table 1.12 UK membership of EC trade lobby organisations.

Organisation	Representing	UK organisations in membership and representatives:
'Groupement'	Millers	NABIM Pat Metaxa (RHM) John Murray (NABIM)
Euromalt	Maltsters	MAGB W.F.S. Mayne (Pauls Malt) P.J. Ellis (Murray Firth Malting Ltd) F. Clive Rattlidge (MAGB)
COCERAL	Grain trade	UKASTA Michael Banks (Sidney C. Banks Ltd) Don Patterson (Dalgety) Ian Douglas (Allied Grain, Scotland) Liz Kerrigan (UKASTA)
		GAFTA David Nelson-Smith (Cargills) Brian Rutherford (BOCM retired) Cliff Haycroft (Soufflet) Maria Capuccio (GAFTA)
	Peas and beans	Charles Bolton (Dalgety) Maria Capuccio (GAFTA) Liz Kerrigan (UKASTA)
FEFAC	Feed compounders	UKASTA Ross Crawford (Carrs) Brian Fawcett (Dalgety) Judith Nelson (UKASTA) Charles Bolton (Dalgety) for pulses
CICILS	Edible pulse trade	BEPA Francis Nichols (John K. King & Son)

COSEMCO	Seed trade	UKASTA
		Michael Dayus (John Bryant (Seeds) Ltd)
		Jim Harley (Alexander Harley Seeds Ltd)
		  for cereal seeds
		Tom Frame (Tom Frame (Seeds) Ltd) for
		  fodder crops
		Laurie Bevin (Booker Seeds Ltd) for
		  vegetable and flower seeds

FEDIOL	Oilseed processors	SCOPA Mark Bennett (BEOCO Ltd)

Abbreviations:
BEPA	Britain Edible Pulse Association
CICILS	Confédération Internationale du Commerce et des Industries des Légumes Secs

Table 1.12 *(contd)*

Abbreviations (Contd)

COCERAL	Comité du Commerce des Céréales et des aliments du bétail de la Communauté Économique Européenne
COSEMCO	Comité des Semence du Marché Commun
FEFAC	Fédération Européenes des Fabricants d'aliments Composés pour Animaux
FEDIOL	European Federation of Seed Crushers and Oil Processors Associations
GAFTA	Grain and Feed Trade Association
'Groupement'	Groupement des Associations Meunières des Pays de la CEE
MAGB	The Maltsters' Association of Great Britain
NABIM	The Incorporated National Association of British & Irish Millers Limited
SCOPA	Seed Crushers and Oil Processors Association
UKASTA	United Kingdom Agricultural Supply Trade Association Ltd

2   The CAP and its Problems

2.1 The CAP price support mechanisms

The CAP was launched in 1962, and at that time general agreement was reached on price support arrangements for cereals. However, the regime as such did not finally come into effect until 1967. The development of the support arrangements for the protein and oilseed crops was still more recent, and was given particular momentum by the need to expand domestic production after the dangers of relying on world market supplies had been highlighted by a threatened US soya export embargo in 1973. Many amendments have been made to the regimes. Nevertheless, the main mechanisms still remain largely unchanged, and are as follows:

(a) The EC acts as a fixed price purchaser – the intervention system. This means that any other purchaser has to make a better offer than is available from intervention. Intervention is thus the floor in the market.
(b) The EC levies imports, which raises the cost of bringing farm products into the European Community, and in the process helps to maintain internal price levels.

Table 2.1 Mechanisms of price support for combinable crops.

Product	Import levies	Intervention	Production subsidy	Other features
Cereals	✓	✓		
Oilseed rape		✓	✓	
Peas, beans & lupins			✓	Minimum price must be paid for processors to qualify for subsidy
Flax & linseed			✓	
Seeds			✓	

(c) It subsidises EC production; such subsidies may be paid directly to the producer, or to the user of the product, to enable them to be able to pay a higher price to the producer.

Different mechanisms apply to different products. Later chapters describe these in detail. They are summarised in Table 2.1.

As the support systems are effective in raising EC prices above world prices, EC exports are uncompetitive unless subsidised. The EC therefore pays export subsidies (restitutions), which in themselves may have the effect of raising EC prices by reducing the availability of supplies for the home market as these are diverted to export.

2.2 Source of the problems

The underlying intention of the EC price support regimes was that, by modifying the supply and demand balance, market prices would be kept higher in times of surplus, and thus more stable than they otherwise would be; but, if intervention purchasing had taken place, the stocks would then be available to supply the market in times of shortage. In practice, however, the various regimes for the major combinable crops have been the main determinant of market prices, and have been the source of significant budgetary cost to the Community. There are three main reasons for this.

First, cereal price support levels were, from the outset, set at a relatively high level to those outside the Community, a decision strongly influenced by German anxieties to maintain a large rural population based on the small farm unit, and to guarantee bread supplies. This in turn has encouraged cereal production, whilst at the same time reducing demand.

Secondly, there has been a rapid increase in cereal yields per hectare. This has been the underlying cause of the expansion of cereal production in the EC (see Fig. 2.1). The Community's cereal area has remained almost constant, with a slight downward trend, though the proportion of wheat has risen. But, if yields continue to increase, the area planted has to fall proportionately for output to remain static (let alone decline). Furthermore, a similar growth in yields has been seen throughout the world. This has caused increasing competition for third country markets, where the EC's emergence as a major cereal exporter has challenged the principal world market supplier, the United States (see Fig. 2.2). This has wider implications discussed later in this chapter.

Thirdly, there have been flaws in the CAP regimes present since their inception. The basic mechanisms of the CAP rely on the application of levies to imported products so that they cannot undercut the prices of home production. But such levies are not applied to cereal substitutes (ingredients which may replace whole cereals in feed), or to proteins and oils. These

EC CEREALS

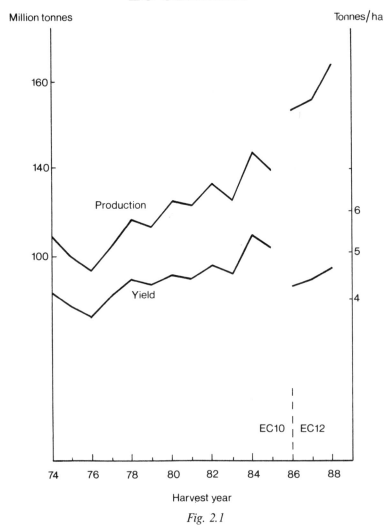

Fig. 2.1

exceptions arise from the arrangements established when the CAP was first set up.

In the GATT negotiations that then took place, the EC accepted reduced rate or zero duty imports of cereal substitutes in exchange for the freedom to operate cereal import levies. It did so because cereal substitutes were thought to be unimportant as world trade levels were insignificant.

In practice, cereal substitute imports expanded and displaced cereals from animal feed rations. This means that huge quantities of EC-produced cereals cannot find a market unless exported. Indeed, the expansion of cereal exports

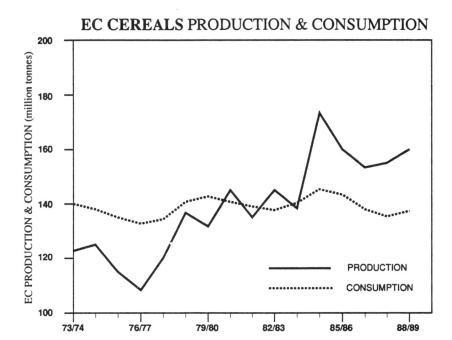

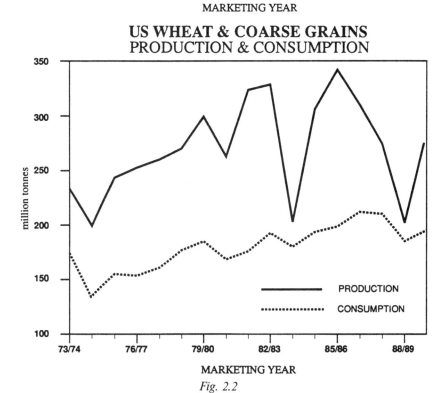

Fig. 2.2

has tended to mirror the increase in imports of substitutes. By 1980/81 the barley equivalent of EC cereal substitute imports had reached 20.7 million tonnes; in the same year 22.7 million tonnes of cereals were exported. The Commission each year estimates the cost to the budget of exporting the cereals displaced by cereal substitute imports. For 1989, this has been estimated as some 1500 million ECU, more than one third of the cereals budget.

Oil and protein crops were excluded from the import levy arrangements because the Community needed to import large quantities with seemingly little prospect of becoming self-sufficient. It was argued that heavy import taxes would raise costs to consumers, whilst the farming industry's inability to expand home production would mean that there would be little off-setting benefit to the rural population.

Again, the reality has been different. Because there are no levies on imported oils and proteins, these are available within the EC at world market prices. The only way the EC can ensure that its growers of oilseed and pulse crops within the EC receive adequate prices is by being prepared to pay production subsidies. In fact, these are paid direct to users, enabling them to purchase EC production at higher prices than would have to be paid on the world market. But such subsidisation is inevitably expensive; in recent years the oilseed rape crushing subsidy has accounted for between one third and two thirds of the market price paid to growers. The budgetary consequences are described in Chapter 8.

2.3 The development of the CAP's problems, and the responses to them

There has been a longstanding awareness of the CAP's difficulties.

By the time the UK joined the EC in 1973, there was an increasing realisation of the problems posed by the rapid expansion of output. Nevertheless, the accession of the UK, Denmark and Ireland provided some respite for the cereals sector; the UK in particular was still a substantial net cereal importer.

But it was in 1973 also that there was a serious threat to Community protein supplies. The US authorities thought that the soya bean harvest there might not be sufficient to meet their national requirements and threatened to stop exports. During that year, mainly as a result of the general shortage of supply, soya meal and other protein prices doubled. It thus became clear that the European Community would have to take steps to boost its own production of proteins (most of which came from processing oil seeds), despite the inevitable budgetary costs. At that time, the EC was only 4% self-sufficient in oil cakes.

By 1980, EC cereal production had moved above self-sufficiency. This, together with the expansion of protein and oil production, meant that a major arable sector cost burden had to be borne by taxpayers, on top of the high expenditure already being incurred to support the milk and livestock industries.

This led in 1981 to the Commission's 'guidelines' document, which stated that it was no longer reasonable to provide unlimited guarantees of price and intervention where there was doubt about the possibility of outlets in the coming years. It also said that Europe's agricultural producers 'must understand that they will have to participate more fully in the cost of disposing of production beyond a certain threshold'.

Further enlargement of the European Community increased its potential financial commitment. The inclusion of Greece added a further million farms, mostly less than ten hectares in size, to the existing 5.5 million farmers in the Community. This, together with a prospect of Spanish and Portuguese accession, gave added impetus for changes to support schemes to avoid Community bankruptcy, though it was recognised that the farming structure could only be changed gradually.

Nevertheless, progress towards reform was relatively slow. Guarantee thresholds had been introduced for cereals and oilseeds in 1982, but the mechanisms were not automatic, and Council decisions were still required. In the end, two events in particular focussed attention on the arable sector.

First, milk, the most costly sector, was brought under quotas in 1984. Secondly, a record cereals harvest that same year prompted alarming production projections, fears of huge stock-piles, and demands to bring cereal production under control. At the same time, oilseeds had also attracted attention, not only because of the increasing cost of support, but also because rape and sunflower oils were seen to be in competition with both butter and olive oil, which were being subjected to radical changes to limit support costs.

1984 was the first year that EC cereal and oilseed prices were reduced. It was also the year in which the UK successfully renegotiated its contribution to the Community budget, and EC expenditure became important to other member states as well.

In 1985, the Council completely failed to agree any prices for cereals and oilseeds. It was left to the Commission to impose price reductions. But in 1986, the Council took two decisions which not only had an immediate effect, but also had a significant influence on further policy developments: it introduced a cereals co-responsibility levy, and the maximum guaranteed quantity system for oilseeds, with price reductions in any year determined by the size of that year's harvest. Both of these decisions were significant in shaping the 1988 stabiliser arrangements.

2.4 The stabilisers

At a Summit meeting in 1988, an important agreement was reached to increase the EC's financial resources. This was approved by all member states, including the UK. It restored the EC's liquidity, but it was part of a package, which meant not only that the CAP was brought under stricter budgetary controls (see Chapter 8), but also imposed stabilisers on most commodities. Full details of the stabilisers applying to cereals, oilseed rape, peas and beans are given in Chapters 3, 4 and 5 respectively.

The imposition of stabilisers was a major watershed for the CAP. The stabilisers introduced the general expectation of continuing price pressure, which could happen without any further decisions by the Council of

Table 2.2 EC Budgets and surpluses. Debate and decisions affecting the arable sector.

1980 – EC 10 becomes net cereal exporter.

1981 – Commission's 'guidelines document' – '... it is no longer reasonable to provide unlimited guarantees of price, and intervention, where there is doubt about the possibility of outlets in the coming years'. The 'threshold' concept introduced the idea of making producers responsible for excess production.

1982 – Council fixes guarantee threshold for cereals and rape seed. Price increases moderated.

1983 – Commission proposes price reduction for cereal and rape seed as guaranteed thresholds exceeded. Price increases 'moderated'.

Commission forward 'CAP adaptation' proposals focussing on price pressure, quality standards, levies and taxes for the cereal and oil seed sectors.

1984 – Introduction of milk quotas shifts focus to control of expenditure in arable sector.

Record cereal crop.

First cereal and rape support price decreases.

Commission acts to reduce expenditure (for example: on bread wheat intervention and carry-over payments).

Fontainebleau agreement provides the UK's budget rebate, shifting the burden of the cost of the CAP.

1985 – Council fails to fix cereal and rape prices.

Commission imposes reduced prices.

Commission's 'green paper' consultation on wide range of options for cereals; conclusions point to realistic price policy, co-responsibility levy, intervention as a safety net, emphasis on quality. Drastic price cuts and quotas rejected. Focus on alternative production, structural and social measures.

1986 – Cereals co-responsibility levy introduced.

Small cereal producer aid introduced.

Intervention period limited; quality standards raised.

Rape seed maximum guaranteed quantity system introduced with automatic same year price cuts.

Budgetary difficulties increase.

1987 – Further tightening of intervention availability and standards.

Buying in at 94% of the intervention price.

1988 – European Council takes decisions which resolve budget problems for the time being and introduce greater budgetary discipline and 'stabilisers'. Funds made available for structural development and set aside.

Ministers. For the main combinable crops it accelerated the trend of declining price support which was already under way, for although the price fixings of recent years have been characterised by compromise, support prices had already been reduced, assisted in many instances by Commission decisions (see Table 2.2 and details in Chapters, 3, 4 and 5).

Stabilisers are intended to limit the costs of price support. Their effectiveness in this will vary according to the mechanisms applied.

If the cereals harvest exceeds a guarantee threshold, an additional co-responsibility levy is charged, and prices are reduced by 3% in the following year. This means that prices can fall repeatedly year after year, but the mechanism is unlikely to stabilise budgetary expenditure in the short term. It will do this as a consequence of prolonged price pressure.

In contrast, the stabilisers for both oilseed rape and pulses may result in unlimited price cuts being applied in the same year as a large harvest. The stabiliser formula means that for each 1% by which the harvest exceeds a maximum guaranteed quantity, prices are decreased by 0.5%. This decrease in prices could well result in the EC saving more than 1% of its aid expenditure per tonne, which would mean, in broad terms, that the larger the harvest the lower the EC's budgetary costs. The exact situation depends on the proportion of the price that is accounted for by EC subsidy. In this case costs are stabilised but the signals to producers are confused as huge price swings may occur from year to year.

The Council of Ministers will have to decide whether the current stabiliser arrangements will continue unchanged for a further period, when the 1988 agreement runs out. The cereals guarantee threshold of 160 million tonnes applies for four years from 1988/89; the oilseed and pulse maximum guaranteed quantities for three years.

2.5 Tackling the problem of cost

Contrary to popular impressions, the greatest problem that has faced the CAP has been that of cost, rather than physical surpluses. Indeed, the EC has never had a stock problem with oil and protein crops, and the cereal intervention stocks that built up after the record 1984 harvest have been disposed of (see Fig. 2.3). EC intervention stocks at the end of the 1988/89 season stood at 6.5% of EC cereal consumption – roughly three weeks' supply: some would term this a buffer stock. It also has to be remembered that one reason for the original build-up was the absence of restrictions on imports of cereal substitutes.

But the cereal sector incurs the major expense of subsidising exports, and costs in the oilseed and proteins sector stem from the expenditure on production subsidies (which cannot be eliminated without stopping subsidised production).

EC INTERVENTION STOCKS
AT THE END OF THE MARKETING YEAR

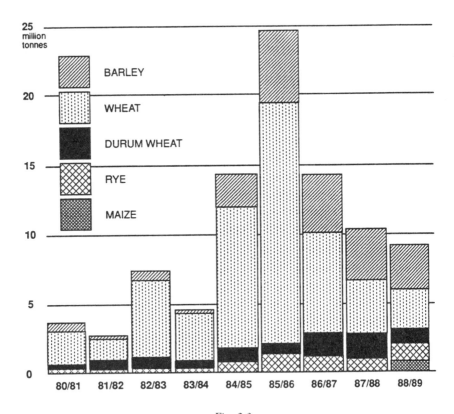

Fig. 2.3

There are a number of ways in which the cost of price support may be reduced. Whilst some may consider that, in the long term, supply and demand will have to be balanced by price, in the shorter term this is the most damaging option for producers. The alternatives under consideration (see Table 2.3) may be seen either as a means of assisting a more gradual adjustment to lower prices, or as long-term solutions. Although the EC is principally relying on cutting producers' prices to control expenditure, it is also making use of other approaches:

Levies

The co-responsibility levy was introduced as a means by which producers

Table 2.3 Major alternative policy options: a summary.

---

1. Co-responsibility levy

For any level of EC expenditure, producer support prices are reduced less by a levy on all production than by lowering intervention prices. When the cereals co-responsibility levy was introduced it was calculated that one-fifth of Community cereal production had to be funded directly from the Community budget, either by intervention purchasing or export subsidisation. In contrast a co-responsibility levy, theoretically at least, applies to all production. It is calculated, therefore, that to have the same budgetary effect, any co-responsibility levy would only have to be one-fifth as large as a reduction in intervention prices.

Problems:

- Protection of producers' prices is a temporary benefit only, if higher prices mean that production continues to expand.
- The yield of the co-responsibility levy is reduced if the amount of production levied (the levy base) is reduced by exemptions or evasion. This may lead to the rate of levy being increased.

2. Quotas

(a) Tonnage quota
Tonnage quota systems usually imply a higher supported price for production in the quota, with production out of the quota being paid for at 'world prices'.

Problems:

- If a higher price is paid by the EC consumer for in-quota production, arrangements must be made to ensure that the over quota excess is exported. The large financial incentive to obtain this cheaper grain is likely to make policing very difficult. All farms would have to be monitored closely.
- If, alternatively, a higher price is paid by direct subsidisation from tax payers, the policy would be too expensive to support prices similar to current levels within the constraints of the EC budget.

(b) Area quota
Allocation of area quotas to cereal production is more practical than tonnage quotas.

Problems:

- Major political problems (freezing the structure of production, and consumer price effects) and, for the UK, the fear of allocation terms being especially favourable to smaller EC producers.

3. Set-aside

(a) Voluntary set-aside
Individual farmers are paid not to grow certain crops on their land and to fallow it or to turn it over to other non-agricultural uses.

Problems:

- The policy is limited in its effectiveness by the money available. It can be made more effective by lowering the overall level of price support.
- Grazed fallow, if permitted, may lead to problems for the livestock sector.
- Steps need to be taken to ensure that the look of the countryside is not ruined.

Table 2.3 (Contd)

(b) Compulsory set-aside
All farmers are required to set-aside a certain amount of land. One option, 'Flexi-quota', envisages that, each year, farmers would have to set-aside an area equal to a certain percentage of their planted cereal area. This percentage could be varied annually.

Problems:

● Policing needs to guarantee that the policy is observed throughout the whole EC.

4. Input limitation

Nitrogen limitation is the most commonly discussed.
(a) Nitrogen taxes

Problems:

● Operates as another form of price pressure.

(b) Nitrogen quotas
Allocation of nitrogen rations to all farmers according to their previous cropping pattern.

Problems:

● Fairness in allocation and policing of quotas throughout the EC.
● Creates inefficiency of production, raising consumer prices.

5. Direct income aids

Generally proposed as an adjunct to price reduction. Special categories of farmers have their incomes protected by direct payments, or a standard payment to all producers for the first 'X' tonnes.

Problems:

● The cost of income support is transferred from consumer prices to the budget, and the scope to maintain incomes is reduced. Price pressure on larger farms may be severe. Protection of smaller EC farms may not take account of UK conditions.

6. Marketing boards

Price pooling by a marketing board arrangement poses similar problems to those for tonnage quotas.

7. Oils and fats tax

An oils and fats tax which would have the effect of raising consumer prices has been proposed as a means of raising funds for the EC budget which could be used to subsidise the difference between EC producer prices and world prices for oil and protein crops.

Problems:

● Strong political lobbying has inhibited progress of this idea. Questions also arise as to its compatibility with GATT.

would contribute directly to the costs of supporting the market. For the short term it is possible to demonstrate that co-responsibility levies are more beneficial to producers than an equivalent intervention price cut. This is because the levy is (theoretically at least) raised on all marketed production, whereas a price cut, in order to achieve the equivalent budgetary saving, would have to be much greater as it applies only to that part of production which is directly financed – i.e. intervention purchases or exports.

Output control

The introduction of a set-aside scheme as part of the 1988 package was a radical new departure for the EC. But leaving land idle is a concept at variance with the philosophy of many Community farmers and farm organisations. As a consequence, set-aside, as a method of limiting the planted area of arable land, has so far not developed into a major policy tool in the EC, in contrast to the USA.

Set-aside can be effective in limiting budgetary cost, and raising producer prices, but to achieve a substantial uptake of set-aside under a voluntary scheme requires either a massive input of financial incentives, or a background of very severe price pressure. The alternative is a compulsory scheme.

For the time being, the other approaches to limiting supply, on-farm quotas or nitrogen limitations, remain theoretical options. However, interest in limiting the use of nitrogen has increased with the EC's concern about nitrates in water.

Increasing consumption

Methods to promote consumption have to be designed so that the existing EC internal market is not undermined. Consequently, the effectiveness of any scheme has to be measured in terms of additional consumption, preferably achieved without loss of returns in existing markets, with the cost balanced against the cost of alternative methods of disposal. Export subsidisation, to enhance EC grain sales on the world market, clearly achieves that aim. The EC also subsidises the manufacture of starch from cereals and other farm products, provided it is used for non-food uses. Other possible major outlets for cereals are increased incorporation in animal feed (to displace imported cereal substitutes), and conversion to bioethanol. Both of these approaches have been pursued recently by the European Community, but the economics remain dubious (see Tables 2.4 and 2.5).

Table 2.4 Cereal incorporation.

As part of the 1988 Summit agreement, the Commission was asked to make proposals for a cereals incorporation scheme. The following proposal was presented that autumn:

Premiums would be paid to encourage the use of cereals in animal feed, as follows:

(a) A progressive premium for additional use exceeding 20% cereals in the ration. Additional use is calculated by reference to that in 1986/7 and 1987/8. The actual scale would be fixed by the Management Committee, but the average rate of premium was estimated at 45 ECU/tonne (then £30).

(b) A general premium of 5 ECU/tonne (£3.50) for cereal incorporation above 45%. This would be paid on the difference between current usage and 45%, or on the difference between usage in the 1986/7 and 1987/8 reference period and 45%, whichever is the lower.

Both premiums could be paid at the same time if the compounder increased an incorporation rate which was already above 45%.

All feed compounders, including livestock producers, who could prove their level of cereal usage in the reference period, would be eligible.

The Commission forecast that the scheme could result in a net saving to the EC of 29 million ECU. This assumed that an aid would be paid on 5 million tonnes of cereals, of which 2 million tonnes would be additional use, thus reducing expenditure on exports.

The proposal raised the important question of how it could be effectively administered, and the danger that in reality expenditure would prove to be greater than the savings, increasing the pressure on the cereals budget. The Council decided against implementing the scheme.

Table 2.5 Bio-ethanol.

The broad make-up of the costs of bioethanol production, without subsidisation, are as follows:

	p/l of ethanol
Wheat (delivered at £95/tonne)	25
Capital and processing costs	10
Expenditure	35
Return from by-product (distillers' grains at £120/tonne)	11
Cost of bio-ethanol	24

The value of bio-ethanol is generally considerably less than this. The largest market is as a fuel, or, importantly now that the EC has committed itself to lead-free petrol, as an alternative to lead as an anti-knock additive in petrol. In this

Table 2.5 *(contd)*

area it is in competition with technically superior oil-derived additives (for example TBA, MBTE, or ETBE). The best price that bio-ethanol might command would be in the region of 80–90% of the ex-refinery price of premium petrol. This price is currently around 10 ppl, the remainder of the pump price being made up of duty, VAT and distribution costs. This suggests a value of bio-ethanol of only 8.5 ppl.

On these figures, the conditions under which bio-ethanol could be produced, and compete successfully in the market as a fuel additive seem to be:

(a) if it benefited from a special waiver of at least 15.5 ppl of the 20 ppl petrol duty;
(b) if the petroleum price rose from 10 ppl to 28 ppl, which would be an increase in crude oil prices from £17–£18/barrel to £50;
(c) if the price of wheat were subsidised down to £32/tonne, delivered.

Or any combination of these changes.

Table 2.6 Imported feed ingredients.

Quantities imported (1987)	Tonnes (thousands)	
	Total	From USA
Soya beans	14,439	10,256
Soya bean meal	10,341	3,109
Manioc	6,986	10
Maize gluten	4,707	4,484
Rice bran	547	24
Brewers wastes	853	757
Fruit waste	1,652	566

Notes:
Soya beans: other suppliers are Brazil, Argentina and other South American countries.
Soya bean meal: Brazil is the largest supplier (4.9 million tonnes in 1987), but its market share is falling with the growth of US exports.
Manioc: Also known as cassava or tapioca, mostly comes from Thailand (5.7 million tonnes in 1987) under quota arrangements (see Table 2.7). Other suppliers are Indonesia and China. More than half is imported via Rotterdam, though significant quantities are transhipped to other member states.
Maize gluten: A by-product of the manufacture of starch and other carbohydrates. Its availability for export from the US has grown with the increase in production of artificial sweeteners, and the development of bio-ethanol production.
Rice bran: Mainly a trade from India to the UK.
Brewers' wastes: Increasing quantities imported, mainly from USA.
Fruit wastes: Mostly citrus pulp, imported mainly from Brazil (1 million tonnes in 1987). Major importers are West Germany and the Netherlands.

2.6 International problems

In considering the way forward for the CAP in the combinable crops sector, the EC's actions have been limited not only by finance and its commitments to the farming population, but also by its international obligations.

Theoretically, the EC may have a wide range of price support mechanisms available to it, but their usefulness is limited by the fact that the EC is open to imports of cereal substitutes, proteins and oils. The main sources of supply are listed in Table 2.6.

The EC is such an important outlet for US farm products – and provides almost the total export market for US cereal substitute products – that the US is determined not to allow any undermining of its access to the Community market. A clear illustration of the US approach occurred when the EC was enlarged to include Spain and Portugal. Under threat of trade retaliation, the EC was forced to agree to preferential terms which would ensure the import into Spain of 3 million tonnes of maize and 0.3 million tonnes of sorghum from the world market. The agreements applied to each year from 1987 to 1990 and talks are being held on extending the arrangements.

The attitude of the US has meant that the discussion of possible changes in the import arrangements for soya, maize gluten, and other products of interest to the US, has had to await the current round of GATT negotiations. In contrast, the EC has had successful negotiations with other suppliers of feedingstuffs.

Table 2.7 Voluntary Restraint Agreements for cereal substitutes.

Manioc
The quantities that can enter the European Community levy-free, or at a reduced rate of levy, are restricted by quotas agreed with the major supplying countries. Other imports are subject to the barley import levy.

The import quotas for Thailand are:

   1983 and 1984: 5.0 million tonnes.
   1985 and 1986: 4.5 million tonnes.
   1987 to 1990 (inclusive): 5.5 million tonnes per annum with a maximum for
   4 years of 21 million tonnes.

In addition 10,000 tonnes of manioc starch may be imported at a duty of 150 ECU/tonne.

Import quotas for China (at 6% duty):

   1986: 300,000 tonnes.
   1987 to 1989 inclusive: 350,000 tonnes per annum.

Sweet potatoes:
Thailand has a quota of 5000 tonnes at a zero rate of duty.

In 1982, following a sharp increase in manioc imports, a Voluntary Restraint Agreement (VRA) was reached with Thailand to limit the quantities imported. Details of this, and more recent agreements with China and Thailand, are given in Table 2.7.

2.7 The General Agreement on Tariffs and Trade (GATT)

(a) GATT and free trade

The General Agreement on Tariffs and Trade (GATT), first signed in 1947, aims to liberalise world trade and thus contribute to economic growth and development. It lays down agreed rules for international trade, applying to some 100 national signatories. From time to time, GATT has been the subject of a number of major re-negotiations which have resulted, amongst other things, in significant world-wide reductions in import duties on industrial products.

(b) GATT and agriculture

When GATT was set up, the US was instrumental in obtaining exceptional treatment for agriculture. In particular, although GATT prohibits export subsidisation, this is allowed for agricultural products, provided it does not result in a country gaining more than an 'equitable share of world trade'.

Despite this, GATT has placed constraints on the European Community which have had considerable implications for arable producers. For example:

(i) When the common external tariff and variable levy system was set up, the EC was obliged to fix (or 'bind') import duties on manioc, maize gluten and citrus pulp. These tariffs cannot be changed without agreement in GATT, and, as noted earlier, the cost to the EC cereals budget has become excessive as the import of these cereal substitute products has increased.

(ii) The Community cannot set up support arrangements which will adversely affect the access to its markets for third country suppliers. This means that it cannot operate support arrangements for oils and proteins which are based on levying imports. Even the EC's present oilseed support arrangements are being challenged by the US, on the basis that their success in raising domestic production is inhibiting the access of US soya to EC markets.

In order to negotiate changes in these arrangements the EC would have to be prepared to pay excessive compensation, beyond that which the Community would be prepared to offer. Instead, such issues are being raised in the framework of the Uruguay Round.

(c) The Uruguay Round

A reduction in agricultural support is now the major focus of the current 'Uruguay Round', which commenced in 1986. Particular aims include:

- improving the competitive environment by increasing discipline on the use of all direct and indirect subsidies and other measures affecting agricultural trade;
- improving import access, in particular by lowering import barriers.

These raise questions of great complexity, and have been the subject of intense debate, particularly between the US and the European Community.

At the start of the Uruguay Round, the US proposed that all agricultural support and protection should be eliminated over 10 years; for 1989 and 1990, there should be a freeze on support and protection.

This was unacceptable to the EC, since it would have implied the dismantling of the CAP. The Community therefore put forward its own proposals, which included the phased reduction of support which directly or indirectly affects trade, and greater use of income support for farmers, particularly those not linked to output.

(d) GATT mid-term review: agreement on agriculture

Agreement was reached on the mid-term review of the GATT 'Uruguay Round' in Geneva in April 1989.

This established that the long-term aim, for 1991 onwards, should be 'substantial progressive reductions in agricultural support and protection sustained over an agreed period of time'. This falls well short of earlier US demands, and is acceptable to the EC because it enables the basic structure of the CAP to be maintained. The first step in this process will have to be taken on the basis of proposals to be tabled by the end of 1989. These will cover both direct subsidies, such as export refunds, and other price support measures which indirectly affect trade.

A number of other issues still have to be decided. In particular, the EC is keen to see 're-balancing', in which reductions in protection on certain products could be offset by increases on others. This is of particular interest to the arable sector in view of the enormous costs which the Community incurs on its protein and oilseed support arrangements as a result of the absence of any protective tariff arrangements. Re-balancing could perhaps allow the introduction of import tariffs for oil and protein crops in exchange for lowered tariff protection for cereals. On the other hand, the US is focussing on 'tariffication': that would entail the conversion of import restrictions and other non-tariff barriers into tariff equivalents. But the Americans have suggested that they would still expect to retain the existing deficiency payment price support arrangements for their farmers, although the EC's support arrangements would be largely dismantled.

It was also decided that in the short term, i.e. up to the end of 1990, domestic and export support and protection levels may not exceed those now in force. In addition, import levies may not be increased, although the CAP's variable levies, such as apply to imported grain, will be allowed to continue provided threshold prices are not raised. During this period support prices to producers should not be raised above present levels. However, for the European Community, this commitment is related to the price in ECU: increases in support in national currencies, for example through devalua-tions of the green pound, will still be permitted. So long as there is a green pound gap, this should be an important benefit to UK producers (see Chapter 8).

All these developments could be highly significant for arable farmers, and may influence the future structure of farm price support. The increased EC emphasis on income aids raises particularly important questions for UK farmers.

2.8 The UK and the CAP

Among the member states, the UK Government has consistently been the strongest voice advocating change and financial restraint for the CAP. This is mainly because the UK became a major net contributor when it joined the Community, at a time when the prosperity of UK citizens compared unfavourably with those of the rest of the European Community, and when the creation of food 'mountains' and higher food prices were becoming evident to the public. This impetus for change has remained, even though the UK contribution to the EC budget has been renegotiated (see Chapter 8).

As the UK Government's priority has been to limit the costs of the CAP, it has seemed that it has sometimes found difficulty in ensuring that its own agricultural industry is treated fairly. Moreover, as the Community has been enlarged, bringing in poorer countries with small farms producing Mediter-ranean crops, there is an increased risk of UK producers' interests being different from those of EC producers generally.

Farmers in the UK are relatively far fewer in number than they are in the rest of Europe (Table 2.8), and the UK is distinguished by a considerably larger average farm size (Table 2.9). This last point is of increasing significance because of the developing pressures in Europe to protect the smaller farmer. For example, when the cereals co-responsibility levy was introduced in 1986, exemption was sought for small producers. This led to

Table 2.8 Proportion of the working population employed full time in agriculture.

UK	2.4%
Belgium	2.8%
Luxembourg	3.7%
Netherlands	4.7%
Denmark	6.5%
Germany	5.2%
France	7.1%
Italy	10.5%
Spain	15.0%
Ireland	15.4%
Portugal	19.0%
Greece	20.0%
EC 12	8.0%
(1987 data)	

Table 2.9 Average size of holdings with arable land.

	Hectares
UK	82.8
Luxembourg	34.9
France	31.7
Denmark	31.3
Ireland	27.2
Germany	17.8
Belgium	16.5
Netherlands	16.0
Italy	6.8
Greece	5.4
EC 10	17.0
(1985 data)	

the introduction of a small cereal producer aid scheme. Notable also, has been the fact that measures to limit the tremendous expense of supporting durum wheat production have focussed on preserving production in the traditional growing areas of the Southern member states, because of the social implications.

3 Cereals

3.1 Outline of producer price support

The possibility of intervention buying of the major cereals (wheat, barley, maize, sorghum and rye) sets a lower limit to grain prices inside the EC. Import levies prevent grain imports undercutting EC prices, and export restitutions, by subsidising exports to third countries, allow EC grain to compete on world markets.

The months when intervention purchasing may take place, and the prices, are set by the Council at the annual price fixing. But, if the European Community harvest the previous year has exceeded a guarantee threshold of 160 million tonnes, the stabiliser mechanism operates to reduce intervention prices by 3 per cent.

Intervention grain is now purchased at a buying price which is set at 94% of the intervention price. This provides a guide to producer price expectations. However, estimations of producers' returns from selling grain to intervention must make allowance for the costs of delivering grain, the delay in payment of nearly four months, and other expenses.

The cereals co-responsibility levy is now a major tax on cereal sales. This levy comprises a basic levy of 3% (of the common wheat intervention price) and an additional levy which varies from year to year. The percentage of additional levy is equivalent to the percentage by which the EC harvest exceeds the guarantee threshold, to a maximum of 3%.

The EC also operates certain special support arrangements, for instance for durum wheat, and provides subsidies for the production of starch for non-food uses.

Set-aside, by limiting output, may also be viewed as part of the mechanism of price support for cereals, as well as for other crops (see Chapter 7).

3.2 Recent changes to EC price support

Although the essential elements of the price support arrangements for cereals have remained largely the same since the launch of the regime, many of the

Table 3.1 Summary of recent changes to EC price support for cereals.

1982	ECU prices increased 8.5%. Guarantee threshold introduced: intended to penalise prices the following year if rolling average of 3 years' production exceeds threshold (after allowing for cereal substitute imports). Bread wheat intervention limited to 3 months.
1983	ECU prices increased 3% (although a price abatement was indicated by the guarantee threshold being exceeded).
1984	ECU prices decreased 1%. Carry-over payments reduced. Bread wheat intervention limited to 3 million tonnes.
1985	Council failed to reach decisions. Commission imposed 1.8% price cut. Delays in payment for intervention.
1986	ECU prices were nominally unchanged but quality standard definitions meant that feed wheat and barley prices were reduced 5%. Full intervention price paid only on wheat meeting bread wheat standards. Premium bread wheat introduced at 2% higher price. Intervention: restricted to September–April inclusive; payment delay 90–120 days (100–150 days for September); monthly increases reduced 4.7%. General intervention standards tightened affecting premiums (e.g. specific weights and moisture content). EC maximum moisture content 15% (UK decides on 14%). Co-responsibility levy introduced: 3% (£3.37) collected from first processors, intervention and export sales. Small producers aid scheme introduced. Start of marketing year changed from 1 August to 1 July.
1987	ECU prices unchanged, but effective support prices reduced as buying-in price reduced to 94% of intervention price; intervention only if triggered by low prices; payment delay for sale to intervention of 110 days; intervention purchasing restricted to October to April with no increments applied between July and October, and size of increment reduced from £1.54 to £1.31 (2 ECU). Intervention moisture content maximum reduced to 14.5% with possible exception of up to 15.5% (UK 15.5%). Carry-over payments discontinued.
1988	ECU prices unchanged but monthly increments reduced 25% to 98p. Intervention operated October to May: triggering mechanism abandoned but buying-in price remains unchanged at 94%. Stabiliser introduced: guarantee threshold set at a 160 million tonnes. Harvest estimated at 162.5 million tonnes. Co-responsibility levy raised to 6%, reduced to 4.6% on 1 January after harvest estimate (rebate paid of £1.647 per tonne). Levy applied at point of first marketing.
1989	3% cut in ECU prices because 1988 harvest exceeded guarantee threshold. Monthly increments reduced 12.5%. Start of intervention delayed until November.

Table 3.2 January intervention buying-in prices for barley.

	ECU price per tonne	Percentage of 83/4 price	Price (£s) per tonne	Percentage of 83/4 price
1983/4	197.43	100	122.14	100
1984/5	195.58	99	121.00	99
1985/6	192.29	97	118.96	97
1986/7	185.17	94	116.10	95
1987/8	166.24	84	109.08	89
1988/9	164.74	83	111.21	91
1989/90	159.37	81	111.78	92

details of the original design have been lost. This is due to the modifications which had to be made as cereal production responded to the price support arrangements, and the Community moved into surplus. In particular, since the Council first decided to apply a guarantee threshold to cereals in 1982, the regime has been constantly adjusted. These changes are summarised in Table 3.1.

Two marked changes to cereal price support – intervention buying operating at only 94% of the intervention price, and the introduction of the cereals co-responsibility levy – were decisions of the Council of Ministers. Certain other important changes have been made by the Commission, rather than the Council. Often these were designed to cut costs, such as delaying payment for grain sold to intervention, and discontinuing carry-over payments for end-of-year stocks; or they raised quality standards.

These, and other attempts at disguised price cuts, have greatly increased the complexity of the EC price support arrangements. As a result, it is difficult to chart the true decline in support prices. Barley has been subject to fewer complicated adjustments of quality and pricing, and so perhaps provides a clearer indication of the change that has taken place. Table 3.2 shows the decline in barley intervention buying-in prices since 1983/4. In addition, producers' returns have been reduced by the co-responsibility levy, and by the effects of inflation.

3.3 Cereal 'stabiliser' or 'guarantee threshold'

(a) The cereal stabiliser mechanism

The European Summit of February 1988 led to the introduction of stabilisers. For cereals, a guarantee threshold of 160 million tonnes applies to each of the European Community harvests from 1988 to 1991 inclusive. If a

Table 3.3 Effect of stabilisers on future cereal intervention buying prices, assuming that the 160 million tonnes guarantee threshold is exceeded each year.

| | Intervention buying-in prices (November) per tonne | |
	Common wheat (£s)	Feed wheat and barley (£s)
1989/90	115.68	109.94
1990/1	112.24	106.67
1991/2	108.90	103.50
1992/3	105.66	100.42

Note:
(1) Assumes no change in green exchange rates, or monthly increments.
(2) Prices before deduction of co-responsibility levy.

harvest exceeds 160 million tonnes, two penalties are applied: an 'additional co-responsibility' levy and a support price cut.

From the beginning of each cereal marketing year, an additional co-responsibility levy, of 3% of the common wheat intervention price, has been collected at the same time as the basic co-responsibility levy. The size of the harvest has then been assessed and a decision taken on any levy refund. The additional co-responsibility levy has been refunded in full if the harvest is judged to be less than the guarantee threshold of 160 million tonnes. If the harvest has been greater than 160 million tonnes, the actual rate of levy applied has been equivalent to the percentage by which the harvest exceeds the guarantee threshold, to a maximum of 3% and the appropriate refund has been paid to producers (see section 3.4C below).

New arrangements will apply from the 1990/91 marketing year. Levy refunds are to be replaced by adjustment of the rate of additional levy in the following year.

If in any year the harvest exceeds 160 million tonnes, then at the start of the next marketing year intervention (and buying-in) prices will be reduced by 3%. These price reductions are cumulative in their effect. Therefore, if the EC harvest exceeds 160 million tonnes in each of the four years from 1988/89, intervention prices will be progressively reduced, as illustrated in Table 3.3, so long as the guarantee threshold remains unchanged.

But it remains possible for EC support prices to be changed by the Council of Ministers, or for national prices to be altered by changes in green exchange rates. As explained in Chapter 8, full devaluation of the green pound should have taken place by the end of 1992.

(b) Exceeding the guarantee threshold

The chances of the EC harvest exceeding 160 million tonnes depend on the

CEREALS GUARANTEE THRESHOLD

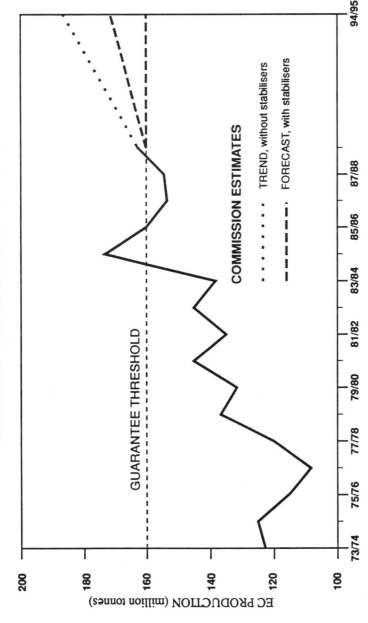

Fig. 3.1

trends in planted area and yields. Figure 3.1 shows the growth of EC production, compared to the guarantee threshold.

As explained in Chapter 2, the most important determinant of EC production has been yield per hectare. The record year of 1984 was characterised by excellent harvests in all the major cereal producing regions of the Community, but the run of unexceptional harvests since 1984 has led to more modest projections of likely yield growth.

The UK Government has taken the view that, in the medium term, the increase in the UK's average yields is unlikely to be much above 1% a year, both for wheat and barley. The rapid growth of average cereal yields in the seventies and early eighties has been attributed largely to a number of factors that will not be repeated, such as the switch to winter wheat, increased use of fertilisers, and the introduction of more managed systems of cereal production, as well as the improvements that are still being made in plant breeding. The UK's situation may be not untypical of many of the cereal-producing regions of the Community.

The planted area of cereals will be liable to vary as a consequence of EC policies, not only on cereals but, perhaps more importantly, on set-aside (see Chapter 7) and the other combinable crops. In particular, the stabilisers for oilseeds and the protein crops may cause significant changes in the areas of those crops planted, with consequent changes in cereals areas.

3.4 Cereals co-responsibility levy

(a) Point of collection

When the cereals co-responsibility levy was first introduced in 1986, the levy was collected on grain which went for processing, export, or sale into intervention. The cost of the levy was passed back to producers.

From 1 July 1988, the principal point of collection of the levy has been the first purchaser of cereals placed on the market by a producer. The purchaser charges the levy back to the producer. Levy is also collected on other grain sold by producers, for instance direct sales to intervention, export, or the futures market, or sales of processed grain (see (g) below).

(b) Rate of levy

The rate of levy collected at the start of the 1989/90 marketing year was 10.44 ECU per tonne, or £7.32. This was 6% of the common wheat intervention price. It comprised two equal parts:

• a basic levy of 3% or £3.66;

- an additional levy of 3% or £3.66 which was later refunded (see below).

Green pound changes after the start of the marketing year do not alter the levy rate.

(c) Adjustment of additional levy

The amount of the additional levy depends on the Commission's estimate of the size of the harvest. In November 1989, the EC harvest was estimated to be 160.5 million tonnes, 0.3% above the guarantee threshold of 160 million tonnes. This implied a levy rate of 0.3% or 38p per tonne, but in the event the EC decided to waive this altogether. Under the arrangements applying in 1989/90 (and previous years) this meant that all the additional levy already paid had to be refunded to producers.

From 1990/91 new arrangements are being introduced whereby there will be no refunds of additional levy, but instead the actual rate of levy collected will depend on the size of the previous year's harvest. The arrangement is that instead of the full 3% additional levy, an additional levy of 1.5% will be collected each year, plus or minus the necessary revision to that rate, depending on whether the actual harvest the previous year was more or less than 160 million tonnes plus 1.5%.

In practice this will mean an additional levy of 1.5% in 1990/91. Then, up to a maximum of 3%, the percentage by which the harvest in 1990 exceeds the 160 million tonne guarantee threshold, will be the percentage rate of additional levy in 1991/2. Similarly, the size of the harvest in 1991 will set the rate of levy in 1992/3 and so on.

(d) Levy liability on grain sales

The levy is payable on all cereals when they are first placed on the market. The definition of cereals includes processed grain.

Where the cereals are first sold as a growing crop, levy is payable when the harvested grain is delivered off the field.

The levy is payable on the weight of grain sold by the producer:

- Where a producer is paid for the actual delivered weight of a consignment, the levy must be paid on that weight, irrespective of the moisture content of the consignment, and of any price deductions that are made.
- Where grain is delivered at a high moisture content and the producer is paid for the dried weight of the grain, levy is payable on the dried weight.
- Where a maximum moisture content is specified in a contract, and the

producer is paid on the delivered weight notionally adjusted to that level, levy should be paid on the adjusted weight.

(e) Exemptions from co-responsibility levy

(i) Cereals used on the farm of origin

Cereals which are not sold but are used on the farm of origin are not liable to levy. Furthermore, provided no sale takes place, this exemption also applies to cereals which may be delivered to a processor for processing under contract, and then returned to the farm, without the physical identity of the cereals necessarily being maintained. But the ration delivered back to the producer must contain grain of the same type and quality as that which he originally supplied to the processor. MAFF field staff operate checks that the rations are consumed on the farm of origin.

(ii) Cereal seed

Cereals delivered under contract to a seed merchant or a seed processor for certification, and which are subsequently sold as seed, are exempt from levy. But, because not all cereals delivered under a seeds contract are ultimately certified and sold as seed, levy is payable on 16% of the seed delivered to merchants or processors. This levy, which merchants sometimes express as a levy rate on all cereal seed deliveries, will be deducted from the price paid to the producer.

Producers may find that their contractual arrangements with merchants make them liable to further deductions equivalent to all, or part of, the co-responsibility levy. Such deductions are not required by EC legislation.

If an individual seed merchant or processor sells as seed less than 75% of the grain delivered to him under seed contracts during the course of the year, he will be required to pay levy on all the cereals for which there is no proof of certification of sale as seed.

(iii) Set-aside

Producers who set-aside at least 30% of their arable land under the set-aside programme may claim reimbursement of the levy paid on marketed cereals up to a maximum of 20 tonnes.

(f) Small cereals producer aid

A small cereals producer aid scheme was introduced in 1986, at the same time as the co-responsibility levy. Each year the UK, and other member states, are allocated a lump sum to finance this aid.

The UK's allocation was £3.35 million and £3.51 million in the first two years, increasing to £6.62 million for 1988/9 when the levy was raised to 6%. The scheme operated allowed all farmers with up to 100 hectares of crops and grass (excluding rough grazing) to reclaim co-responsibility levy on a maximum of 25 tonnes of grain. The number of claims has consistently fallen short of expectations, and each year the UK has only been able to pay out about a half of the total available sum.

Those claiming aid in respect of the 1986/7 and 1987/8 years were paid in full (up to £84 and £88 respectively) and claims for the 1988/9 year are also expected to be met in full (up to £135.33).

For the 1989/90 year, the UK's allocation of funds has been cut to £3.05 million. This may mean that applications exceed the funds available in which case claims will have to be scaled down. Revised rules are to be operated. Claims will only be allowed from producers with 80 hectares or less of crops and grass (excluding rough grazing).

The maximum payment for 1989/90 (after allowing for the general rebate of the additional co-responsibility levy) is expected to be £91.50. This will be reduced by any necessary scaling down according to the availability of funds.

Applications for aid have normally been required by 31 July and payment has been made by the end of the year.

(g) Registration and collection of levy

Cereal purchasers are required to register with the Intervention Board. In addition, growers have to register with the Intervention Board if they sell their own grain in processed form, such as in animal feed, if they sell grain direct outside the UK, or if they sell direct to the Futures Market.

Payment of levy collected on transactions within a three-month period must normally be made by the end of the month following this period. In the case of those who expect to collect levy on less than 250 tonnes per year, annual payment of levy may be authorised. Interest is charged on amounts which are not received by the due date.

The Home-Grown Cereals Authority acts as agent for the Intervention Board in collecting the co-responsibility levy in the UK.

(h) Advice

Enquiries concerning the operation of the levy should be addressed to: The Cereals Co-responsibility Levy Unit, The Intervention Board, PO Box 69, Reading RG1 7QW (Tel: 0734 583626).

MAFF Divisional Offices handle all claims for levy repayment (including small producer cereal aid, and set-aside exemptions).

3.5 Cereal intervention

(a) UK intervention purchases

(i) Prices

For the marketing year 1 July 1989 to 30 June 1990, intervention purchasing may take place for the months of November to May inclusive. Buying-in prices (which are 94% of intervention prices) are shown in Table 3.4. These prices rise by monthly increments.

Grain will be accepted for intervention provided it meets minimum standards (see Table 3.5). The additional standards that have to be met for wheat to qualify for the full common wheat price, or the premium wheat price, are given in Table 3.6. Quoted intervention buying prices refer to grain of standard quality. Price adjustments for quality differences are applied as set out in Table 3.7.

(ii) Offers to intervention

In the UK intervention is operated on behalf of the Intervention Board by the Home-Grown Cereals Authority (H-GCA). Applications to sell grain to intervention are made to the H-GCA's Regional Cereals Officer (RCO) from whom the appropriate forms may be obtained. These have to be submitted to the RCO for the area in which the grain is located. Offers must be made, and received, between 1 November and 31 May.

The minimum amount that can be offered for intervention is 100 tonnes,

Table 3.4 Intervention buying-in prices (1989/90).

	Feed wheat, barley, rye (£s/tonne)	Common wheat (£s/tonne)	Premium wheat (£s/tonne)
July–October*	109.02	114.76	117.05
November	109.94	115.68	117.97
December	110.86	116.60	118.89
January	111.78	117.52	119.81
February	112.70	118.44	120.73
March	113.62	119.35	121.65
April	114.54	120.27	122.57
May	115.46	121.19	123.49
June*	109.02	114.76	117.05

* Intervention does not operate in these months.

Table 3.5 Cereal intervention minimum quality standards.

To be acceptable for intervention, grain must be sound, fair and of marketable quality. It shall be regarded as such if of typical colour, free from abnormal smell and live pests (including mites) at every stage of their development, and meeting the requirements set out below.

	Wheat	Barley	Rye
Maximum moisture content (%)	14.5	14.5	14.5
Minimum specific weight kg/hl	72	63	68
Maximum total impurities (%)	12	12	12
of which:			
(a) broken grains	5	5	5
(b) grain impurities	12	12	5
of which:			
other cereals, grains damaged by pests and grains with discoloured germs	5	5	—
grains overheated during drying	3	3	3
(c) sprouted grains	6	6	6
(d) miscellaneous impurities	3	3	3
of which:			
noxious seeds	0.1	0.1	0.1
ergot	0.05	—	0.05

The temperature of the grain must not exceed 15°C (18°C in November), and it must conform to maximum residue limits for certain pesticides.

Table 3.6 Additional standards for common wheat and premium wheat.

For wheat to be accepted as common wheat, or premium wheat, and qualify for the full price premium, the grain has to meet the standards in Table 3.5 and the following higher standards.

Physical criteria:

Maximum total impurities	10%
Maximum grains damaged by spontaneous heating and excessive drying	0.05%
Maximum grains damaged by heat during drying	0.5%

Technical criteria:

	Premium wheat	Common wheat
Hagberg Falling Number (min)	240	220
Zeleny Index (min)	35	20
Dough Machinability	Pass	Pass
Protein content (% dry matter, min.)	14.0	11.5

The discounts which apply to lower standards are given in Table 3.7 below.

Table 3.7 Adjustments to intervention prices for variations in quality.

Intervention buying-in prices are adjusted for variations in quality by amounts determined as a percentage of the buying-in price at the start of the marketing year (see Table 3.4). For wheat, the price adjustments are all based on the common wheat intervention price.

Moisture content:
Premiums are payable on grain with a moisture content of 13.4% or lower. For each 0.1% by which the moisture content is below 13.5%, a 0.1% bonus applies, but below 10% moisture there is no additional bonus.

Specific weight:
Price deductions apply for low specific weights, *viz.*:

Wheat:

Below 76.0 – 75.0 kg/hl	0.5%
Below 75.0 – 74.0 kg/hl	1.0%
Below 74.0 – 73.0 kg/hl	1.5%
Below 73.0 – 72.0 kg/hl	2.0%

Barley:

Below 64.0 – 63.0 kg/hl	1.0%

Rye:

Below 70.0 – 69.0 kg/hl	0.5%
Below 69.0 – 68.0 kg/hl	1.0%

Admixture and impurities:
Price deductions apply as follows for each 0.1% above the standard quality:

Broken grains, above 3%	0.05%
Grain impurities (including shrivelled grains)	
barley and wheat, above 5%	0.05%
rye, above 3%	0.05%
Sprouted grains, above 2.5%	0.05%
Miscellaneous impurities, above 1%	0.1%

Technical testing:
Grain offered for technical testing but which fails to meet the full standards for common wheat is subject to the following price reductions, which are not cumulative. The maximum penalty for failing any of the tests is 5%.

Hagberg Falling Number	below 220	5%
Zeleny Index	below 20	5%
Dough Machinability test	fail	5%
Protein content	below 11.5 – 11.0	1%
	below 11.0 – 10.5	2%
	below 10.5 – 10.0	3%
	below 10.0 – 9.5	4%
	below 9.5	5%

or 500 tonnes in the case of wheat submitted for technical testing. The offer may be made up of grain in up to four places.

(iii) Technical testing

Technical testing (for Hagberg Falling Number, Zeleny Index, Protein Content, and Dough Machinability) is necessary for wheat offered as common wheat or premium wheat (but not for feed grains). The offeror has to pay the test fee, of £312 plus VAT, or if the grain is in two, three or four locations, respectively, £390, £468, and £564 plus VAT.

The grain is sampled, at the point of offer, by a contracted representative

UK CEREALS INTERVENTION CENTRES

1. ABERDEEN	20. LEITH
2. ASHFORD	21. LIVERPOOL
3. AVONMOUTH	22. LOCHARBRIGGS
4. BELFAST	23. LONDONDERRY
5. BERWICK-UPON-TWEED	
6. BRIGHTON	24. MANBY
7. CAMBRIDGE	25. MONTROSE
8. COWES	26. NEWCASTLE
9. DARLINGTON	27. NORTHAMPTON
10. DONCASTER	28. NORWICH
11. DRIFFIELD	29. OLD DALBY
12. ELGIN	30. OXFORD
13. EXETER	31. PERTH
14. GLASGOW	32. PLYMOUTH
15. HARTLEBURY	33. PREES HEATH
16. INVERGORDON	34. RIPON
17. INVERNESS	35. SOUTHAMPTON
18. IPSWICH	36. TILBURY
19. KINGS LYNN	37. TURRIFF

*All centres accept Barley and Wheat;
Cambridge is the only centre accepting Rye.*

Fig. 3.2

of the Intervention Board. The sample is split: one part is given to the offeror to retain, one part is retained by the RCO, and one part is sent to the Flour Milling and Baking Research Association, where the tests are carried out.

(iv) Delivery and intake

When an offer to intervention is accepted (which in the case of common wheat and premium wheat is after the H-GCA receive satisfactory technical test results), the offeror is given delivery instructions specifying the store location and delivery dates. Any proposed changes in arrangements have to be referred to the H-GCA Regional Cereals Officer.

The offeror has to make the arrangements and pay for delivery. However, if the grain has to be transported to a store which is further away than the nearest intervention centre, the offeror is paid a transport allowance of 3 pence per tonne per excess mile. If the distance is less, a deduction is applied at the same rate. Figure 3.2 lists the intervention centres.

On intake, the storekeeper weighs and samples the load. The offeror has the right to witness, check and dispute the sampling. The sample is divided: one part is kept by the offeror, one for reference at the store, and the third part is tested by the storekeeper.

(v) Payment

The offeror completes a claim form, stating the tonnage delivered, and the mileages to the intervention centre and to the intervention store. If his claimed tonnage matches the storekeeper's intake tonnage for the contract, he is sent a contract record sheet, which shows all the storekeeper's test results, and a 'pricing document' showing the make-up of the total amount payable.

The price paid is normally that for the month of delivery allocated by the H-GCA. Any grain delivered earlier than this is paid for at the price for the earlier month. At the end of the season, deliveries allocated, and made, in June, are paid for at the price of the month of offer. The storekeeper's analysis, and the results of any technical tests, provide the basis for calculating price adjustments.

The Intervention Board makes payment between the 110th and 115th day following that on which the final load was delivered. (The same time-scale applies whether or not the final load was one that was rejected.)

(vi) Penalties and appeals

● *Failure to complete the contract:*
An offer to intervention becomes binding when delivery instructions have been received by the offeror, although exceptions to this rule apply if there is

a delay in the offer being accepted (of 10 working days or more after receipt of the offer and completion of any necessary technical tests), or if the time allocated for delivery to store commences more than one month after the instructions are issued.

Failure to comply with the full contract conditions (either in time or quantity) will lead to penalties, generally £3 per tonne for shortfall against the contract conditions, plus £60. However, a tolerance of plus or minus 25 tonnes may be allowed without penalty (provided the delivered tonnage is still above the minimum quantity for offer to intervention). But quantities delivered in excess of the plus 25 tonne tolerance, and the total deliveries if less than 80 tonnes have been delivered, have to be removed from the store at the offeror's cost.

● *Dispute of technical test results:*
An offeror may ask for a re-test if he disputes the technical test results. The grain will not be re-sampled, but the offeror will have to pay a further test fee of £50 + VAT, which will be refunded if the appeal succeeds.

● *Rejected loads:*
Any load failing to meet minimum quality standards when delivered to store is rejected. The offeror is charged a rejection fee of £12 a load.

The storekeeper's findings regarding infestation, smell and temperature are final. Appeals against rejection on other grounds may be made to the Regional Cereals Officer, provided the load can be kept separately and under the storekeeper's control pending the appeal result. If the dispute is unsuccessful, the offeror has to pay the cost of handling, storage and testing by an independent analyst.

● *Dispute of physical test results:*
Appeals against the storekeeper's test results may be made to the Regional Cereals Officer. The offeror has to pay a fee of £50, refundable if the appeal is upheld.

(b) Intervention prices and market prices

Whilst intervention price support still underpins producer cereal prices, as the system is adjusted to become more of a market of last resort, so the price becomes less of a guide to producer price expectations.

The early season market in the UK, and other northerly parts of the EC, is not directly supported except by the anticipation of intervention opening in November, and the operation of intervention buying in the southern member states of the Community from August. Predictions of harvest prices have to take account not only of the immediate supply and demand situation, but also of the true cost of storage and financing grain for the early months of the marketing year. With high interest rates in the UK, this is likely to be markedly above the 92 pence per tonne increment built into the intervention system.

Actual producer returns from sale into intervention are significantly below intervention buying prices. Mainly this is because producers bear the cost of transport to the intervention store (say £4 to £6 per tonne), and payment is made 80 days later than for a normal commercial transaction (cost say £3 per tonne with 14% interest rates). In addition producers have to pay for technical testing in the case of offers of bread-making wheat. Individual offerors will weigh the balance of other factors such as bonuses for drier grain, the possible discounts on other criteria, and the costs associated with grain rejection.

Of course, producers may benefit from prices higher than the ex-farm price implied by intervention buying prices. This may be because the quality offered is exactly what the market needs, but it may also be because supplies are limited as the result of heavy intervention buying or export sales. In such situations, shortages may result, particularly at the end of the season. The Commission controls the release of grain from intervention, and, if grain is in store but cannot be released, a technical shortage will raise prices. But, equally, such a shortage may disappear and market prices fall if the decision is taken to release stocks onto the home market.

The EC operates intervention buying for all the main cereals, but certain grains are not purchased, such as triticale and oats. The assumption is made that the markets for these grains will be supported by the general buoyancy of the remainder of the cereals market, as well as by the cereal import protection arrangements, in which they are included. Also, under EC rules, intervention in a particular region is only operated if there is likely to be a surplus. For this reason, the UK does not offer intervention for durum wheat or maize.

Intervention provides no direct support for grain which cannot be conditioned to meet the quality requirements. Nevertheless, if such supplies are relatively limited, their prices may still benefit from the strength of the market derived from the operation of the intervention system. In previous years, derogations have sometimes been granted to reduce intervention standards if harvest conditions are particularly bad, such that there is a large volume of grain below intervention quality. But the current approach of making the intervention system less attractive may greatly reduce the likelihood of any derogations, except perhaps for moisture content standards.

(c) Sale of intervention grain

(i) Tenders

The principal method of selling intervention grain, either for the home market or for export, is by tender.

Tenders for sales of large quantities of grain from intervention to the home market are only operated if the Commission considers that this is essential from the point of view of the EC market; a Management Committee decision which will take into account the availability of supplies in the whole of the Community.

The Intervention Board may open a tender for the sale of up to 999 tonnes, without needing Commission approval.

(ii) Prices

Grain sold from intervention is subject to minimum price rules designed to ensure that it does not undercut the market, and create a 'carousel' of grain being sold into and out of intervention.

Bids for intervention grain for the home market must be at or above whichever is the greater of:

- the buying-in price for the month of the tender or
- the local market price at the place of storage.

In the UK, the local market price is determined by discounting by 1% (to allow for prepayment of grain purchased from intervention) the H-GCA average price for the county in which the store is located. Prices are taken from the H-GCA Weekly Bulletin published on the day prior to the bids being considered. If there is no county price shown in that issue, the regional price is used instead; failing that, the UK price.

(iii) Invitation to tender

Details of the grain available for sale and the tender arrangements are available from the Intervention Board.

Table 3.8 Storage categories for UK intervention wheat.

Type 'A'	Hagberg at least 240. Zeleny at least 35. Protein at least 14% (dry matter basis). Machinability test: pass.
Type 'B'	Hagberg at least 220. Zeleny at least 20. Machinability test: pass.
Type 'C'	Hagberg at least 220. Zeleny at least 20. Protein at least 11% (dry matter basis). Machinability test: fail.
Type 'D'	Wheat not falling into the above categories, and all other wheat not offered for technical testing.

The grain which is offered for sale will be at named intervention stores, and the quality at each store is specified by the average intake quality. Since the 1986/7 marketing year, intervention wheat has been stored under four separate categories, according to quality; these are listed in Table 3.8.

The tender document will specify the premiums or deductions which apply to purchases from particular stores. These are based on the average quality of the grain on intake into the store. The actual price that a successful purchaser has to pay for the grain is the price that he bids, plus or minus these quality adjustments.

(iv) Making a bid

Bids, which must be made within the tender period, are submitted to the H-GCA together with a security of 5 ECU per tonne (£3.51). The tenderer specifies the store, the grain he is bidding for, the quantity (which must be at least 20 tonnes) and the price.

The bids are expressed as ECU per tonne (see Chapter 8 for conversion factor) and must comply with the minimum price rules explained earlier. The bid relates to grain of standard quality, and the offeror must bear in mind that the amount actually paid for the grain may be different because of the quality premiums or deductions explained above.

The tenders are normally adjudicated each Tuesday, and the tenderer notified of the outcome on Tuesday or Wednesday. Securities are released in respect of unsuccessful bids.

(v) Payment and collection of grain

Successful bids have to be paid for within one calendar month of the date of contract (which is the date a successful bid is notified), or the security is forfeited.

Arrangements for discharge of grain are made in conjunction with the storekeeper. The grain will be stored at the Intervention Board's cost for one month after the contract date.

The Intervention Board has to be paid in full for the grain before collection.

3.6 Import levies and export refunds

All cereals imports and exports have to be licensed. In addition, third country exports usually qualify for refunds (restitutions) and imports are subject to levies.

Two types of export refund are available. The standing rates of export refund, which are fixed by the Commission at least once a month, are always available, and may vary according to destination, and the product being exported. Higher refunds may be obtained by tendering; such bids are considered by the Commission, and Cereals Management Committee, each week. Exporters may fix all refunds in advance.

In order to decide on the level of export restitutions, a comparison is made between the price of grain (FOB) in the EC, notably at Rouen, with the price of competing North American grain (FOB Gulf). Typically, decisions on export restitutions also take into account the EC internal market and budgetary situation as well as international trading conditions.

Import levies are imposed at a level which should ensure that cereals imported from outside the European Community cannot be offered on the Community internal market at a price below the Target Price. In general terms the latter is the price of a cereal considered to be reasonable for the EC's deficit regions and is related to the intervention price. Specifically, target prices are derived from intervention prices by adding a 'market component' and the cost of transport between Ormes (in France) and Duisburg (in West Germany) regarded as the respective centres of cereals surplus and cereal deficit. The threshold price is the price applicable at the EC external frontier and is derived from the target price after allowing for unloading and transport costs. The Commission calculates import levies daily as the difference between the threshold price and the lowest price (generally at Rotterdam) at which grain of an appropriate quality is available

IMPORT LEVIES AND EXPORT REFUNDS

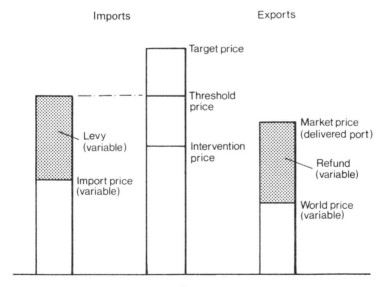

Fig. 3.3

for import into the Community. These price relationships are described in
Figure 3.3.

3.7 Other cereal price support

(a) Durum wheat aid

For 1989/90, the intervention buying-in price of durum wheat was 238.06
ECU or £166.97 per tonne. Cumulative monthly increments of £1.25 apply
from November to May.

Producers in traditional durum-producing regions of the south of the
European Community benefit from a producer subsidy of 158.98 ECU per
hectare (80.61 ECU in Spain).

(b) Starch

The EC pays a refund for starch production, provided the starch is used for
non-food purposes (textiles, pharmaceuticals, paper, adhesive etc.). The
same rate of aid is paid for production from wheat, rice and potatoes. For the
quarter beginning 1 July 1989, this was 86.5 ECU per tonne.

(c) Flint maize

A production aid of 155 ECU per hectare was introduced in 1989 for high
quality flint maize, in order to encourage the EC to produce the maize needed
for cornflakes and displace imports. The aid is restricted to areas which can
achieve a 15% moisture content by natural drying pre-harvest.

(d) Maize, breadmaking rye and sorghum intervention

In those countries which operate intervention for maize, breadmaking rye
and sorghum, the buying prices are the same as the barley price (155.44 ECU)
for sorghum, and the same as the common wheat price (163.62 ECU) for
breadmaking rye and maize.

(e) Rice

The EC offers intervention for rice (314.19 ECU per tonne) and also an aid
of 330 ECU per hectare to encourage the production of Indica rice.

4 Oilseed Rape

4.1 Outline of producer price support

Market prices paid to producers of rapeseed are maintained substantially above the price of imported rapeseed by the payment of a crushing subsidy. The crushing subsidy is available to oilseed crushers who use rapeseed produced within the European Community. The rate of aid is meant to reflect the difference between the world market price of rapeseed and the EC rapeseed target price. In practice, the rate of crushing subsidy has lately been somewhat lower, but pitched at a level sufficient to maintain the market price paid to producers above the rapeseed intervention buying price.

Intervention is available for rapeseed from November to May. The buying-in price is 94% of the intervention price.

Double zero rapeseed (also called double low) has low levels of both erucic acid and glucosinolates, in contrast to single zero (single low) rapeseed which only has a low level of erucic acid. The meal left after the oil has been extracted is more valuable as animal feed than the meal obtained from single zero seed. Double zero rapeseed commands a premium, both for aid payments and for intervention. Single zero rapeseed is being phased out as an EC-supported crop.

The Council sets the target and intervention prices at the annual price fixing, but these are subject to stabiliser adjustments, decided by the end of August each year. There is no limit to the extent of the price cut which may be imposed as a result of the stabiliser; it depends on the estimated size of the harvest. Provisional rates of aid are operated for the two months before the stabiliser price adjustment has been agreed. Intervention and target prices are increased by monthly increments, designed to encourage an even flow of rapeseed onto the market.

4.2 Recent changes to EC price support

The expansion of production of EC oilseeds has achieved the target of improved Community self-sufficiency in vegetable oils and proteins, but this has placed a heavy burden on the EC budget (see Chapter 8). By 1987 each tonne of rapeseed was costing £200 or more to subsidise; the EC was financing more than two-thirds of the producer price. This situation gave added impetus to changes, in both price support and quality standards, which were already under way.

The Commission's concern with the quality of rapeseed started with the problem of erucic acid in rape oil. This led to the setting of a single zero standard, but also sensitised the Commission's thinking such that it reacted rapidly to the demonstration of the potential hazards to animal health of glucosinolates in rape meal. The double zero standard was introduced, at first only qualifying for an intervention price bonus; later a premium on the crushing subsidy as well. Now double zero is forecast to be virtually the only type of rapeseed which will be supported from the 1992/3 marketing year onwards.

A switch to double zeros was expected to expand the market for rapemeal, but, when first mooted, it also presented the possibility of cutting EC rapeseed support costs. Double zero rapes were almost only grown in Denmark, where they were spring-sown and yielded markedly less than the autumn-sown varieties cultivated in the rest of the EC. Growing rape under these conditions throughout the EC would decrease average yields, and also make the crop uneconomic for many farmers. This would be a saving for the EC.

However, autumn-sown varieties which are low in glucosinolates, and which have acceptable yields, have since been developed. But whether the switch to double zero does, in fact, take place without a long-term cut back in production depends crucially on the level of glucosinolate content which will be acceptable as 'double zero'.

The problem of the budgetary cost of rapeseed support has also been tackled directly. In 1981 a guarantee threshold was introduced, and prices in ECUs were decreased for the first time in 1984. Later, in 1986, rapeseed became the first arable crop to be subject to a maximum guaranteed quantity with automatic price cuts. The stabiliser agreement of 1988 removed any limits on the potential price cuts that could result from an increase in production.

The chronology of the changes in EC support for oilseed rape is shown in Table 4.1.

4.3 Oilseed rape stabiliser

The European Summit of February 1988 set a new rapeseed maximum guaranteed quantity for the EC (excluding Spain and Portugal) of 4.5 million tonnes. See Fig. 4.1. This applies for the three marketing years 1988/89 to 1990/91. Separate MGQs were set for Spain and Portugal.

Each year, before 31 August, the size of the harvest is estimated. If this

Table 4.1 Summary of recent changes to EC support for oilseed rape.

1981

Guarantee threshold agreed, starting at 2 million tonnes for 1981/2.

Price decrease in following year if rolling 3 year average of production exceeds
production threshold.

Intervention premium introduced for double zeros.

1983

For the first time prices are influenced by the guarantee threshold, and the
increase is limited to 4% (instead of 5.5%).

Commission proposes lower prices for single zeros.

Commission makes specific proposals for oils and fats tax (in view of enlargement
of EC and potential costs of olive oil support).

1984

Prices decreased 2%.

1985

Price decrease of 1.8% (for exceeding threshold) imposed by Commission after
Council fails to agree prices.

1986

Guarantee threshold replaced by maximum guaranteed quantity (MGQ) of 3.5
million tonnes with price effect in same year. Harvest estimated before start of
marketing year.

No price cut triggered, but monthly increments decreased.

Double zero bonus of 12.5 ECU per tonne applied to both target price and
intervention price.

Warning that from 1991 support will be limited to double zero varieties.

1987

Prices: cut 3%,
            plus further cut of 10% of target price for effect of MGQ,
            plus reduction of intervention buying prices to 94% of intervention
            prices.

Intervention: delayed until October,
                increments decreased 20% and delayed to start November.

Double zero bonus increased to 25 ECU per tonne.

1988

MGQ set at 4.5 million tonnes. Provisional aid rates until end August.

Stabiliser calculation raises buying-in prices by 2.9% compared with previous
year.

Increments decreased by 20%.

1989

Council decision that from the start of the 1992/3 marketing year aid will be
restricted to double zeros and high erucic acid rapeseed intended for industrial
use.

Start of intervention delayed to November.

Monthly increments decreased 12.5%.

Stabiliser raises buying-in prices by 5.2% compared with previous year.

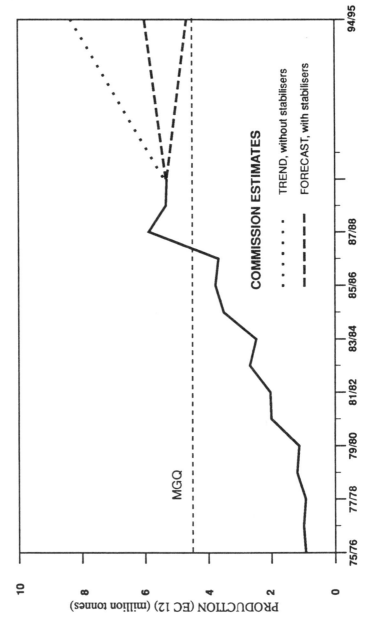

Fig. 4.1

estimate exceeds the maximum guaranteed quantity, the prices in the current
year are reduced by:

0.45% for each 1% overshoot in 1988/89;
0.5% for each 1% overshoot in later years.

The price reduction is calculated by reference to the 1987/88 target price
(before it was abated by 10% because of the operation of the maximum
guaranteed quantity mechanism in that year), i.e. 450.2 ECUs (now worth
£315.76 at a green rate of 1 ECU = £0.701383).

This price reduction, in money terms, is subtracted from the intervention
buying-in price for rapeseed. (Prior to 1989, the price reduction was
subtracted from the intervention price.)

After the end of the marketing year, a further estimate is made of the size
of the preceding year's crop. If this differs from the earlier (31 August)
estimate, the support prices applied to the new crop will be adjusted (either
up or down) to allow for the error in the calculation of the previous year's
support prices.

Provisional rates of aid are operated for the first two months of the
marketing year, whilst the effect of the 31 August harvest estimate is awaited.

The calculation of the stabiliser price adjustments for 1989/90 are shown
in Table 4.2, with the resulting intervention buying-in prices in Table 4.3.

In order to be able to take a view of next year's prices the stabiliser
adjustment has to be predicted. For this two estimates are needed:

- First an assessment of the final view that the Commission will take of the
  previous year's harvest. Each 0.1 million tonnes difference between the
  Commission's first August estimate of the harvest and its estimate at the
  end of the season can change the 1990/91 price by 1%. This poses the
  question as to how good the Commission's August estimate was.
- Secondly, a view of the size of the current year's harvest, which means
  assessing both EC plantings and yield expectations.

The intervention buying-in prices are then calculated as follows:

Step 1. Adjustment for previous year's harvest estimate:

$$\frac{\text{August harvest estimate (mn te)} - \text{end year estimate (mn te)}}{4.5 \text{ mn te}} \times 100 \times 0.5 \times 450.2 \text{ ECU}$$

This price adjustment may increase or decrease prices but there is no positive
adjustment for the extent to which the harvest may fall below the MGQ.

Step 2. Current year adjustment:

$$\frac{\text{August harvest estimate (mn te)} - 4.5 \text{ (mn te)}}{4.5 \text{ mn te}} \times 100 \times 0.5 \times 450.2 \text{ ECU}$$

Step 3. Add (1) and (2) and convert to sterling using the green rate.

Step 4. Deduct (3) from single zero intervention buying-in price before stabiliser adjustment (at August 1989, this was 383.14 ECU or £268.73). Add monthly increments, and double zero premium, if appropriate.

Table 4.2 The stabiliser calculation.

The harvest estimates used for the August 1989 stabiliser calculation were:

    1988 (final):   5.2 million tonnes
    1989 (forecast):   4.9 million tonnes

The calculation was as follows:

(1) Correction for final estimate of 1988 harvest
At August 1988, the harvest was estimated at 5.3 million tonnes. This was 17% higher than the maximum guaranteed quantity (MGQ) of 4.5 million tonnes. In that year a price cut of 0.45% was applied for each percentage point of overshoot; the resulting price cut was 7.65% of the 1987/8 theoretical target price before the application of stabilisers (450.2 ECU); a cut of 34.4 ECU.

But the 1988 final harvest estimate (made at August 1989) was 5.2 million tonnes. This was only 15% higher than the MGQ, which meant that a price cut of only 6.7% (i.e. 15% × 0.45), or 30.4 ECU, should have applied, instead of the 34.4 ECU cut that was used. The credit of 4 ECU was carried forward to the 1989/90 prices.

(2) Effect of 1989 harvest estimate
The 1989 harvest estimate at 4.9 million tonnes was 8% above the MGQ (this calculation ignores the figures after the decimal point). For 1989, there is a price cut of 0.5% for every 1% by which the MGQ is exceeded. This resulted in a 4% price cut, which is 18 ECU.

(3) The 1989/90 prices
The 1989 cut of 18 ECU, and 1988 credit of 4 ECU gave a net price cut of 14 ECU. This was subtracted from the July theoretical 1987/8 buying-in price of 383.14 ECU, which applied before the stabilisers were introduced. The addition of the monthly increments, and conversion to sterling at the green rate of exchange (1 ECU = £0.701383), gave rise to the prices shown in Table 4.3.

The figures underlying the Commission's August 1989 estimate of the 1989 harvest were:

- Plantings: 1989 EC rapeseed area of 1.652 million hectares (10% down on 1988). (The UK area was 9% down at 323,000 ha.)
- Yields: 1989 EC average yield of 2.97 tonne/hectare (up 5.7% on 1988). The figure used for the UK was 3.00 tonnes per hectare (2.99 in 1988).

Table 4.3 Rapeseed intervention buying-in prices (£s/tonne).

| | Single Zero | | Double Zero | |
	1988/9	1989/90	1988/9	1989/90
July–October*	230.18	258.91	247.71	276.44
November	232.27	260.86	249.80	278.39
December	234.36	262.81	251.89	280.34
January	243.26	264.76	260.79	282.29
February	245.41	266.71	262.94	284.24
March	247.56	268.66	265.09	286.19
April	249.70	270.61	267.23	288.14
May	251.85	272.56	269.38	290.09
June*	230.18	258.91	247.71	276.44

* Intervention offers are not accepted in these months.

4.4 Oilseed rape crushing subsidy

The crushing subsidy is designed to enable oilseed processors to crush seed produced in the European Community in competition with seed imported from third countries. The aid is supposed to match the difference between the target price, which rises by monthly increments between November and May, and the world market price. It is fixed at least once a week. Double zero rapeseed receives a crushing subsidy premium of 25 ECU (£17.53).

The world market price for rapeseed, which is a key factor in calculating the aid, is determined weekly by the Commission on the basis of the most favourable purchasing opportunities, and may be adjusted to take account of the prices of competing products. Rape oil and meal prices may be used to derive a rapeseed price, after allowing for the costs of crushing. It is assumed that one tonne of seed yields 0.39 tonnes of oil and 0.56 tonnes of meal, the remainder being moisture and impurities.

Since world market prices experienced in Community countries will vary according to the strength of the currency, the crushing subsidy is adjusted by monetary differential amounts (MDAs) (see Chapter 8).

Crushing subsidy and MDAs can be pre-fixed up to five months in advance. This reduces the effect of fluctuations between the time of purchase of seed and the processing being carried out. In a period when the rates of aid on offer are considered to be advantageous, this may signal the pre-fixing of a large tonnage of aid. In turn, market prices may benefit.

Subsidy is only paid on rapeseed of Community origin. Before processing, the delivered rapeseed has to be identified and placed under control at the mill. Oil, moisture and impurities content is recorded, and the mill has to process within 150 days. The subsidy is paid on proof of processing, or in advance on payment of a security.

4.5 Oilseed rape intervention

For the marketing year 1 July 1989 to 30 June 1990, intervention purchasing of oilseed rape operates for the months of November to May inclusive. The buying-in prices, after the effects of stabilisers have been taken into account, are shown in Table 4.3.

As for cereals, rapeseed intervention is administered by the H-GCA for the Intervention Board. There are Intervention Centres at Cambridge, Hull, Leith, Liverpool, Oxford and Tilbury. The arrangements for selling rapeseed to intervention are similar to those for cereals, described in Chapter 3, but there are two important differences:

● All rapeseed offers are sampled at the point of offer. The sample is sent to ADAS at Cambridge for testing for oil content, erucic acid and moisture. Then, for double zero offers, the sample is sent to Wolverhampton for glucosinolate testing. This can be a somewhat lengthy procedure. (For double zero offers a test fee of £45+VAT is payable by the offeror.)

● Offers made at the end of the season, which are delivered to store in a month with a lower price than the month of offer, are paid for at that lower price.

The minimum quantity of rapeseed which may be offered is 100 tonnes. The minimum quality is set out in Table 4.4.

Table 4.4 Rapeseed intervention: minimum quality.

The rapeseed must be sound, fair and of marketable quality and shall be regarded as such if of typical colour, free from abnormal smell and live pests (including mites) at every stage of their development and meeting the following requirements:

  Maximum moisture content: 9%
  Maximum total impurities: 2%
  Maximum erucic acid content: 5%

The rapeseed must not contain yellow millet or canary seed, or show any other evidence of having been denatured. It must not contain more than 2% of seeds which are crushed, damaged, sprouted, split, skinned, broken, overheated, immature, empty or mouldy. Immature seeds are those which are not black or dark brown and which when cut open with tweezers or a scalpal have green flesh; if the flesh is yellow they are counted as mature seeds.

Double zero rapeseed: To qualify for the double zero bonus, the seed must also have a glucosinolate content of not more than 35 micromoles per gram at 9% moisture.

Table 4.5 Rapeseed intervention: adjustment to standard quality.

Standard quality for rapeseed:

Moisture 9%
Impurities 2%
Oil content 40%

Adjusted weight:
Payment for rapeseed is on the basis of an adjusted weight.
The delivered weight is converted to an adjusted weight by multiplying by:

$$\frac{100 - (\text{sample impurities \% + moisture \%})}{100 - (2\% + 9\%)}$$

However if the moisture content of the sample is below 6%, the figure of 6% is used in the formula instead of the actual moisture percentage.

Oil content:
A premium or deduction is applied of 0.2 ECU per tonne for each 0.1% oil content above or below the standard quality.
The oil content assessment is adjusted to take account of the change to adjusted weight. The oil content used as the basis of payment is the oil content percentage of the delivered sample multiplied by:

$$\frac{\text{delivered weight}}{\text{adjusted weight}}$$

Standard quality is defined as seed with a moisture content of 9%, an impurities content of 2% and an oil content of 40%. The actual price paid is adjusted up or down according to the extent that the seed is above or below standard quality. The formulae for making these adjustments is shown in Table 4.5.

To qualify for the double zero price bonus of 25 ECU/tonne, in addition, the rapeseed must have a glucosinolate content of not more than 35 micromoles per gram of seed at 9% moisture.

4.6 Target prices, intervention prices and producer prices

The detail of the calculation of crushing subsidies provides scope for the Commission to exercise its own judgement about the necessary level of subsidy. The theory of producers attaining the target price has not generally been borne out in reality. Instead, intervention buying-in prices provide the guide for producer price expectations (though allowance has to be made for different conditions of sale). The Commission needs to fix rates of crushing subsidy at a level which will generally maintain market prices higher than intervention buying-in prices, if the EC is to avoid the costs of carrying substantial intervention stocks.

The operation of the maximum guaranteed quantity can result in major reductions in support prices. However, in a year with a low harvest in addition to higher support prices, the demand for rapeseed can lead to a significant market premium.

The more widespread production of double zero varieties in the UK highlighted the problem for producers of achieving a premium for double zero rapeseed. The extra aid paid to crushers does not have to be passed on to the producer. The producer's premium for supplying double zero rapeseed has to be secured in the sales contract.

4.7 Double zero rapeseed

Double zero rapeseed has a low content of both erucic acid and glucosinolates. A high erucic acid content is unacceptable in rape oil for food use, and the normal EC standard specifies a maximum content of 5% in the seed. High glucosinolates limit the extent to which rape meal can be incorporated into animal feed.

The EC standard for the glucosinolate content of double zero rapeseed is as follows:

Current standard: 35 micromoles per gram of seed.
From 1 July 1991: 20 micromoles per gram of seed.

At present, seed meeting the double zero standard qualifies for a premium over the single zero price.

From the start of the 1992/93 marketing year, aid will be restricted to double zero rapeseed, and only those single zero varieties which are high in erucic acid, intended for industrial use. The latter have to be further defined.

The definition of the acceptable content of glucosinolates for double zero rapeseed has been the subject of lengthy debate. The discussion has been about both the level of glucosinolates and the method of measurement. The various analytical methods measure different glucosinolates, and so the measured levels of glucosinolates vary with the method adopted. Differences between glucosinolates raise the question as to whether some are more toxic than others, but the necessary research to determine this has yet to be carried out.

Ideally, the most toxic glucosinolates need to be the ones which are being measured in the standard method of glucosinolate testing. If this is not the case, the benefits of switching to double zeros will be diminished, and the efforts of plant breeders may be misdirected.

This uncertainty over the scientific basis for defining glucosinolates complicates the debate on the question of an acceptable EC standard. There is a current EC reference method for testing glucosinolates, known by the abbreviation TPGLC (temperature programmed gas-liquid chromatography). However, scientists continue to debate whether this reference method should be changed.

Because the reference method requires sophisticated laboratory techniques, and is costly and somewhat lengthy to carry out, the EC also approves a number of quicker testing methods.

Those approved for use in the UK are X-ray fluorescence (XRF), and a glucose release method developed by Unilever (the Colworth method). The X-ray fluorescence method is regarded as quick and reliable, but the equipment is expensive. The Colworth method has the advantage of speed without requiring expensive equipment.

4.8 Sunflower seed

The mechanisms of price support applying to sunflower seed are the same as those which apply to oilseed rape. However, before the operation of MGQs, sunflower benefits from generally higher prices and monthly increments (respectively, 534.7 ECU per tonne, compared with 407.6 per tonne for rapeseed, and 3.31 ECU per tonne compared with 2.78 per tonne for rapeseed). It has a separate maximum guaranteed quantity of 2 million tonnes (for the Community of 10). The 1989 sunflower harvest was estimated to be 2.313 million tonnes, cutting nominal intervention prices by 36 ECU to under 500 ECU per tonne, or £350 per tonne. Sunflower seed is considered to be a more desirable product than rapeseed; neither oil nor meal pose any quality problems.

4.9 Soya

EC production of soya has been expanding rapidly. The 1989 harvest has been estimated at 1.8 million tonnes, compared with only 200,000 tonnes in 1984.

The EC support arrangements for soya are similar to those for peas and beans (see Chapter 5). Prices are limited by the operation of a maximum guaranteed quantity of 1.3 million tonnes.

Before the effect of the maximum guaranteed quantity, the 1989/90 minimum price payable to producers was 489.4 ECU per tonne (with a lower price in Spain). The maximum guaranteed quantity calculation reduced this by 19.3 per cent of the guide price, or 107.8 ECU per tonne, to 381.6 ECU (£268) per tonne.

5 Pulses (Peas, Beans and Lupins)

5.1 The price support mechanisms

The EC provides subsidies to animal feed compounders using peas, beans and sweet lupins in order to encourage the purchase of EC production instead of imported proteins. To ensure that the benefit of the EC subsidy is passed back to producers, aid is only paid to the compounder if in turn the producer has been paid at least a minimum price, adjusted by the appropriate number of monthly increments.

The subsidy is set at a level that should make EC-produced pulses purchased at the minimum price competitive with imported soya and maize gluten.

The Council of Ministers sets minimum prices for peas, beans and lupins at the annual price fixing, but these prices are subject to the operation of a maximum guaranteed quantity. Final support prices are unknown until the end of August.

Aid is also given for dried peas and beans used for human consumption, but this is calculated on a different basis. It is fixed as the difference between the Guide Price, set by the Council of Ministers at the price fixing, and the world market price of peas or beans (adjusted for quality standards). The aid is altered if world market prices change substantially.

The aid schemes are based upon standard qualities of 14% moisture and 3% admixture (and, in the case of lupins, 3% bitter lupins).

5.2 Recent changes to EC price support

The EC's support scheme for protein crops has been highly successful in encouraging an expansion of production of peas and beans. Since the introduction of the aid scheme for compounding peas and beans, output has increased substantially, feed peas by more than ten times. In order to avoid any market distortion from aiding peas and beans for one outlet only, the scheme was later widened to include peas and beans for human consumption. It was then extended to lupins, in order to encourage this potential source of proteins.

Table 5.1 Summary of changes to EC support for pulses.

1978	Aids introduced for compounding peas and beans.
1982	Aids introduced for dried peas and beans for human consumption.
1984	Scheme extended to include lupins.
	1% reduction in minimum prices.
1985	Monthly increments introduced for peas and beans (September to February) but:
	average bean price reduced 3.5% and July price cut 5.4% (£9.59);
	average pea price unchanged but July price cut 1.9% (£3.40).
1986	Increase in minimum price of peas, beans and lupins: EC 1%, UK 2.3%.
	Monthly increment added for March and April.
1987	Cut in minimum and guide prices: EC 10%, UK 5.8%.
	MDAs introduced for peas and beans (negative effect on UK prices).
1988	Stabiliser introduced: minimum prices cut by 10.3% for peas and 10.7% for beans (£17.43).
1989	Price review cuts bean prices 4%.
	Monthly increments for peas and beans cut 12.5%.
	No significant change in prices as a result of the stabilisers.

But the scheme has been costly, because of the absence of any protection against imports. Also, whilst the transfer of land from cereals to pulses has the benefit of moving from the production of a surplus crops to one in deficit, the expansion of peas and beans may also displace some of the demand for cereals.

The future expansion of pulses will now be curtailed by the imposition of a maximum guaranteed quantity.

Table 5.1 outlines the changes that have been made to the price support for peas and beans.

5.3 The pulses stabiliser, or maximum guaranteed quantity

The support prices fixed by the Council of Ministers are subject to adjustment by the operation of the maximum guaranteed quantity mechanism.

In 1988, as part of the stabiliser arrangements, a maximum guaranteed quantity (MGQ) of 3.5 million tonnes was set for peas, beans and lupins, applying to each marketing year from 1988/9 to 1990/91.

If the Community harvest (estimated before the end of August each year) exceeds the maximum guaranteed quantity, price reductions are calculated as follows:

• For 1988/9: 0.45 per cent of the guide price for each 1 per cent production above the MGQ.

Table 5.2 The stabiliser calculation.

The harvest estimates for EC production of peas, beans and lupins used in the August 1989 stabiliser calculation were:

1988 (final): 4.276 million tonnes;
1989 (forecast): 4.060 million tonnes.

The stabiliser calculation was as follows:

(1) Correction for final estimate of 1988 harvest
At August 1988, the harvest was estimated at 4.2 million tonnes, 20% higher than the MGQ of 3.5 million tonnes. Minimum prices were cut by an amount equal to 0.45% of the guide price (295.2 ECU) for each 1% by which the MGQ had been exceeded; a cut of 9%, or 26.6 ECU.

As the final estimate was higher, the actual percentage by which the MGQ is exceeded is 22%, and the price cut should have been 9.9% or 29.2 ECU.

The additional price cut of 2.6 ECU was carried forward to reduce 1989/90 prices.

(2) Effect of 1989 harvest estimate
The 1989 harvest estimated at 4.060 million tonnes was 16% above the MGQ of 3.5 million tonnes. For 1989/90 the price reduction which applied was 0.5% for each 1% by which the MGQ was exceeded. This amounted to a cut of 8% or 23.6 ECU.

(3) 1989/90 prices
The two price cuts, calculated above, were added to give a total price reduction of 26.2 ECU, or £18.38. This amount was deducted from the July minimum prices (which after the last price fixing and before the operation of stabilisers, were: peas £180.75, beans £163.12), and the appropriate monthly increments were added, to derive the schedule of prices given in Table 5.3.

The statistics for the pulse stabiliser are complicated by the involvement of more than one crop, and by the fact that the figures used are the estimates of the quantities claiming aid. These figures are given above. Actual production estimates are about 13.5% higher. The underlying statistics were:

1989 EC yields
Peas: 4.07 te/ha, down 8% on 4.42 te/ha in 1988.
Beans: 2.27 te/ha, down 18% on 2.76 te/ha in 1988.

1989 EC area
Peas: 896,000 ha, up 13% on 792,000 ha in 1988.
Beans: 424,000 ha, down 6% on 453,000 ha in 1988.

Lupins
1988 production: 6000 tonnes.
1989 production: 10,000 tonnes.

Table 5.3 Peas and beans and lupin minimum prices (£/tonne).

	Beans		Peas	
	1988/9	1989/90	1988/9	1989/90
July	145.66	149.04	151.64	162.37
August	145.66	149.04	151.64	162.37
September	146.85	150.15	152.82	163.48
October	148.03	151.26	154.00	164.59
November	149.21	152.37	155.18	165.69
December	150.39	153.48	156.36	166.80
January	151.57	154.58	157.54	167.91
February	152.75	155.69	158.72	169.02
March	153.93	156.80	159.90	170.13
April	155.11	157.91	161.08	171.24
May	155.11	157.91	161.08	171.24
June	155.11	157.91	161.08	171.24

Lupin minimum prices: 1988/9: £170.34. 1989/90: £182.36.

- For 1989/90 and 1990/91: 0.5 per cent of the guide price for each 1 per cent of production above the MGQ.

The minimum price and the rates of aid are reduced by the same amount of reduction (in £s or ECUs) as applies to the guide price.

At the end of the marketing year, the final harvest estimates are made. If the final estimate would have resulted in a different MGQ price reduction, the MGQ in the following year is adjusted.

The mechanism works in a very similar way to that for oilseed rape described in Chapter 4.

Table 5.2 shows the calculation of the MGQ stabiliser adjustment, and derivation of minimum prices for 1989/90.

The minimum prices applying in 1989/90, after the MGQ adjustment has been made, are shown in Table 5.3. The guide prices are shown in Table 5.4.

The methods used for estimating the crop are of importance in gauging the accuracy, or inaccuracy, of the MGQ decisions, and the likelihood of substantial corrections applying because of differences between August and end year estimates. The later estimates used in the MGQ calculation are based upon aid claims. The forecasts of the crop, which are made in August of the year of the harvest, are based on early August crop surveys (by ADAS in England and Wales). The total amount of production estimated by this method is reduced to allow for on-farm usage without aid, and for seed production. This allowance is calculated as 0.22 tonnes multiplied by the EC total planted area.

Table 5.4 Peas and beans guide prices (£/tonne).

	Beans		Peas	
	1988/9	1989/90	1988/9	1989/90
July	176.24	188.67	176.24	188.67
August	176.24	188.67	176.24	188.67
September	177.42	189.78	177.24	189.78
October	178.60	190.89	178.60	190.89
November	179.78	192.00	179.78	192.00
December	180.96	193.11	180.96	193.11
January	187.42	194.22	187.42	194.22
February	188.64	195.33	188.64	195.33
March	189.86	196.44	189.86	196.44
April	191.08	197.55	191.08	197.55
May	191.08	197.55	191.08	197.55
June	191.08	197.55	191.08	197.55

Early season marketing of pulses in advance of the August settlement of the MGQ requires special contract arrangements which allow for the fact that the minimum price, and the aid rates, are unknown.

5.4 The minimum price

(a) Payment of the minimum price

To qualify for aid, the peas, beans or sweet lupins must be the subject of a contract between the grower and first buyer. The price paid must be at least equal to the minimum price based on standard quality, with increments appropriate to the month of delivery to the first buyer.

Standard quality is 14% moisture and 3% impurities.

The minimum price is paid on an adjusted weight. This is calculated according to the moisture and impurities content of the sample. Table 5.5 sets out the method of calculation of adjusted weights, and the additional reductions which apply in the case of high moisture contents.

Products have to be 'sound, genuine and merchantable', though deliveries need not be at standard quality in order to be acceptable. If the quality delivered is unacceptable and needs to be cleaned and/or dried in order to become usable, the scheme states that the costs of drying and cleaning have to be agreed between the producer and first buyer, and paid by the producer. These costs have to be shown separately. If a first buyer has to remove part of a delivery because it is not 'sound, genuine and merchantable', the price

Table 5.5 Calculation of adjusted weight.

The quantity of products delivered is multiplied by:

$$\frac{100 - (\text{sample moisture \% + impurities \%})}{100 - (14\% + 3\%)}$$

In those cases where moisture contents are above 17% the weight calculated by the above formula is further reduced by deducting:

(a) For deliveries between 17.01% and 22.99% moisture:

   0.006 × delivered weight × the number of percentage points the moisture content is above 17%

(b) For deliveries at or above 23% moisture two deductions apply:

   0.034 × actual weight, and then

   0.004 × actual weight × the number of percentage points the moisture content is above 23%.

---

paid to the producer may be reduced to allow for the costs incurred, including loss of weight.

Under certain circumstances the relevant date for determining which minimum price is payable to the producer (and the rate of aid paid to the processor) may not be the actual date of delivery. If, after drying, the producer is paid the minimum price (with drying costs separately itemised), then the post-drying payment date may be used as the change of ownership and delivery date. Similar rules apply if change of ownership is delayed beyond the delivery date and a storage charge is made. These circumstances may allow a higher minimum price to be paid.

(b) The minimum price and producers' prices

If peas and beans are purchased below the minimum price, they are ineligible for the EC subsidy. In the past, when markets have been difficult it was often reported that market movement has been dependent upon trade taking place at prices effectively below the minimum price, albeit that this had to be disguised.

But the absence of any guaranteed outlet, such as intervention, creates a fundamental problem for growers, who may feel forced to take a lower market price than the minimum. This situation is likely to reflect the fact that the subsidy has been fixed at an inadequate level so that peas and beans are not competitive with other proteins available.

Under these circumstances, sticking to the letter of the regulation and having no sales below the minimum price could mean that growers would be severely penalised by the rule that is intended to protect them.

From 1 July 1990, including new forward contracts from 1 January 1990, new EC regulations will apply aiming to try to prevent sales below the minimum price. Aid will only be paid if the pulses have been purchased by an approved first buyer; if an approved first buyer fails to pay the minimum price, he will be required to pay compensation to the producer which is equal to at least twice the difference between the minimum price and the actual price paid.

5.5 Claiming aid

(a) Marketed peas and beans

For all purchases of peas and beans, a first buyer must obtain a First Buyer Certificate, which is necessary for the eventual aid claim. This requires compliance with the rules regarding contracts with growers, sampling and testing of deliveries, payment of the minimum price, and submission of a delivery declaration. A requirement of the delivery declaration, showing the date, weight and price paid, is that it has to be signed by the grower. But once a First Buyer Certificate has been obtained, it does not have to remain with a particular lot; it may be used in support of any claim (for the same species).

End users purchasing pulses have to give advance notice of delivery to MAFF and submit an entry declaration. The pulses have to be weighed, sampled and tested for moisture and impurities (and also for bitter grains in the case of lupins). They may be officially sampled and tested. MAFF will issue a certificate of identification, quoting the tonnage and the rate of aid (which may have been pre-fixed). The pulses have to be used for an approved product within seven months of identification and a declaration of use submitted to MAFF.

Final payment of aid is made once the claim has been substantiated against the first buyer's records. An advance payment (made not more than 14 days after the entry declaration) may be paid if the end-user lodges a security equal to the aid. The security is released once the claim has been verified.

The aid rate payable will depend upon the end product. The human consumption aid rates apply to peas and beans prepared for food use.

First buyers are required to be registered, and end-users have to be registered and approved. All have to keep appropriate records. The scheme requires close attention to paperwork and checks and controls because of the heavy losses that can be incurred from failure to receive the aid, or from the penalties that are applied for failure to comply with the required timetable.

The details of the scheme and the arrangements needed to cope with the requirements of large purchasers, and the continuous processing of modern mills, make the scheme generally complex to operate. And, of course, in some cases end-users are also first purchasers.

(b) Producer groups

Special arrangements allow the payment of aid to approved producer groups using pulses for animal feed. Pulses delivered to the producer group have to be notified in advance to MAFF, weighed and sampled, then milled for animal feed or denatured by colouring or addition of fish oil.

Approved producer groups have to include at least 30 producers of peas, field beans or sweet lupins, and be able to guarantee the use of the products only for their own livestock. The full amount of aid has to be passed on to the producers, and transfer prices have to comply with minimum price requirements. Detailed rules govern record keeping and accounting.

Applications for approval as a producer group should be addressed to MAFF.

(c) On-farm compounders

Farmers can claim aid if they grow peas, field beans or sweet lupins, and process them for animal feed compounds. However, the regulations must be strictly adhered to. Full details can be obtained from the Intervention Board.

The farmer must create separate legal entities (such as a company or partnership) so that he can operate separately as a 'first buyer and end user' as well as being a farmer. These separate legal identities will have to have different names and separate bank accounts.

The farmer must apply for registration as a first buyer, and apply for approval as an end user of peas and beans.

Strict records will have to be kept including separate stock records for home-produced and purchased products, with details of the weight of product taken in (with moisture and impurities contents), movement of products between stores, quantities processed, invoices and documentation for products bought and sold.

The processing machinery must be permanently fixed.

The first buyer enterprise enters into a contract with the farmer to buy pulses at the minimum price (including appropriate monthly increments). On delivery, pulses have to be sampled for moisture and impurities. A delivery declaration is lodged with the Intervention Board, and a certificate is issued which will support a claim for aid once the products have been processed. As soon as processing has taken place, the Intervention Board has to be notified, the certificates submitted, and the aid will be paid as soon as operations have been verified by the Intervention Board's inspecting officers.

5.6 Fixing aid rates

At the annual price fixing, the Council of Ministers, as well as setting a guide price for peas and beans, and minimum prices for peas, beans and lupins, fixes an 'activating threshold price' (or activating price) for soya meal. The activating price is related to the minimum price through the assumption that the feeding value of peas and beans is equivalent to a mix of 55% cereal and 45% soya meal (40% and 60% respectively, in the case of lupins).

The Commission fixes the rate of aid for compounding peas, beans and lupins at least twice a month. It bases the rate of aid for compounders on 45% of the difference between the world market price of soya meal (assessed on the basis of the most favourable Rotterdam purchasing possibilities) and the activating price. But the Commission also takes into account prices of 23% protein maize gluten, and world price trends for soya and competing products.

Aid for peas and beans for human consumption is set as the difference between the guide price and the world market price. It is set at the start of the marketing year, and changed in response to substantial fluctuation in world market prices.

Aid rates, as well as minimum prices, are affected by the operation of the maximum guaranteed quantity, as explained earlier.

Monetary Differential Amounts are applied to rates of aid. They are intended to adjust aid rates to compensate for differences between green and market rates of exchange (see Chapter 8).

As well as current rates of aid, the Commission publishes a series of six projected rates for subsequent months. These aid rates may be advance fixed by end-users. (Applications for advance fixing have to be supported by a security.)

5.7 Vetches, chick peas and lentils

In 1989 the EC introduced a new support scheme for grain vetches, lentils and chick peas, based on an aid payment per hectare. This may be subject to a stabiliser reduction in any year when the area exceeds the maximum guaranteed quantity calculated as the average area planted in the EC in the three marketing years 1985 to 1988. The scheme was expected to be of particular interest in Spain and Portugal.

5.8 Dried fodder

Since 1974 the EC has operated price supports for dehydrated and dried fodder, as another protein source. A production aid is paid per tonne. Main uptake has been in France and Italy.

6 Seeds and Linseed

(A) SEEDS

6.1 EC Directives

Production and marketing standards for seeds are laid down by EC directives, with a view to protecting the interests of seed purchasers. The directives, dating originally from 1966 to 1970, detail the arrangements for seeds sold in the EC. Only varieties listed in the Common Catalogue may be marketed. To be included, varieties have to be tested and proved satisfactory. Individual seed lots of those varieties may only be marketed if officially certified as having met the necessary field inspection and seed standards. The seed then has to be labelled according to official standards.

The EC also provides seed production aid. This is now mainly restricted to herbage seeds. It is argued that the seed production of most other crops may be underpinned by the appropriate commodity support arrangements.

6.2 EC Common Catalogue and National Lists

The Common Catalogue lists those varieties of plants whose seed may be generally marketed in the Community. The Common Catalogue is compiled from the National Lists of each member state.

To comply with the EC Directive, each member state is required to maintain a national list of varieties officially accepted for certification and marketing in its own territory. Only varieties which are *distinct*, *stable* and *sufficiently uniform* (D.U.S.) may be accepted. Varieties must also be of satisfactory *value* for *cultivation* and *use* (V.C.U.), except in the case of grasses not intended for fodder production.

A variety is satisfactory for cultivation or use if, at least in a particular region, by comparison with other varieties in the National List, its qualities, taken as a whole, offer a clear improvement either for cultivation, or as regards the uses which can be made of the crops or products derived from it.

Where other superior characteristics are present, individual inferior characteristics may be disregarded.

Member states also have a duty to ensure that varieties on their National Lists are properly maintained.

Whilst a variety must be listed in the Common Catalogue in order to be generally marketed, such listing also means that no marketing restrictions may be applied from 1 January following two clear years after a variety has first been listed. Exceptions to this rule allow a member state to prohibit a certain variety on the basis of lack of distinctness, stability or uniformity, or where its cultivation may be harmful from the point of view of plant health, or because of lack of value for cultivation and use in the member state concerned. The variety may still be accepted in member states where such problems do not occur.

Varieties from third countries may be included in the Common Catalogue under special arrangements. The procedures for official examination of varieties, and their maintenance, operated by the country concerned have to be established to be equivalent to EC standards.

Normally, a variety remains on the Common Catalogue for 10 years, or longer if it is still in widespread use, but it may be removed if it is proved that it is no longer distinct, stable, or sufficiently uniform. Marketing of such seed then ceases after three years at most.

Responsibility for adding varieties to the National List rests with MAFF but the technical work on which such decisions are based is done by the National Institute of Agricultural Botany (NIAB) in England and Wales, the Scottish agricultural colleges, and the Department of Agriculture for Northern Ireland (DANI). These organisations also produce the Recommended Lists of the best varieties on the National List. The Common Catalogue arrangements mean that varieties may be marketed in the UK which have not been included in the National List trials.

6.3 Seed certification

The EC directives determine that marketed seed has to be produced to certain specific standards, and officially certified. The standards required are more stringent for seed generations closer to the stock maintained by the breeder.

The generations are named according to the following sequence: breeder's seed, pre-basic, basic, certified first generation (C1), certified second generation (C2). In general the directives specify standards for basic and certified seed. The category 'commercial seed' is also used in the case of some fodder crops.

In addition to the minimum standards set by the directives, the UK operates a Higher Voluntary Standard (HVS) for cereals and some herbage seed crops.

The production of certified seed requires inspection of the seed crop in the field, as well as inspection, after harvest, of samples of the seed prepared for sale. This is done by the Official Seed Testing Station (OSTS, which is part of NIAB) and its private licensed testing stations.

Table 6.1 Cereal seed: crop standards. Varietal purity, species purity and wild oats.

Crops to produce	Level	Varietal purity (percentage by number)	Species purity (percentage by number)	Wild oats: maximum number per hectare			
				Oats	Barley	Wheat	Rye
Basic Seed	Higher Voluntary Standard (2)	99.95	99.99	Nil	7	7	—
	Minimum standard	99.9	No standard	Nil	7	7	7
Certified Seed	—	(1)	—	—	—	—	50
Certified Seed 1st Generation	Higher Voluntary Standard (2)	99.9	99.99	Nil	7	7	—
	Minimum standard	99.7	No standard	Nil	20	50	—
Certified Seed 2nd Generation	Higher Voluntary Standard (2)	99.7	99.99	Nil	7	7	—
	Minimum standard	99.0	No standard	Nil	20	50	—

Notes:
(1) For rye the number of plants of the same crop species which are recognisable as obviously not being true to the variety shall not exceed one per 30 sq m for the production of Basic Seed and one per 10 sq m for the production of Certified Seed.
(2) Crops to produce seeds at the Higher Voluntary Standard must not be more than one-third laid at the time of inspection.

The field inspection standards include verification of:

- previous cropping;
- isolation distances;
- varietal purity;
- species purity;
- injurious weeds;
- disease status.

The seed sample is tested for:

- germination;
- analytical purity;
- moisture content.

Extracts from the standards applying to various crops are given in tables as follows:

Cereals: Tables 6.1, 6.2 and 6.3;
Oilseeds: Tables 6.4 and 6.5;
Peas and Beans: Tables 6.6 and 6.7.

For complete details, reference should be made to the appropriate seed regulations, last amended September 1989.

If temporary difficulties occur in the supply of basic or certified seed, authority may be given to market seed subject to less stringent quality standards, or varieties not included in the Common Catalogue or national lists. Such derogations are only granted for limited time periods, and only if the seed shortage cannot be overcome within the Community. Requests for derogations are referred by MAFF to the EC's Standing Committee on Seeds and Propagating Material for Agriculture, Horticulture and Forestry. Its decisions are taken following Management Committee procedures (see Chapter 1).

Table 6.2 Cereal seed: barley and wheat. Crop standards for loose smut infection.

	Minimum Standard	Higher Voluntary Standard
	(maximum percentage by number)	
Basic Seed	0.5	0.1
Certified Seed of the First Generation	0.5	0.2
Certified Seed of the Second Generation	0.5	0.2

Seeds produced from a crop which has failed on official examination to meet these standards may nevertheless be eligible for official certification if they have been adequately treated by any approved method for the control of loose smut, or if an embryo test carried out by an official seed testing station, shows that the seeds meet those standards.

Table 6.3 Seed standards: cereals.

| | WHEAT, BARLEY AND OATS | | | | | | RYE | |
| | Basic | | Certified 1st Gen. | | Certified 2nd Gen. | | | |
	Min.	HVS	Min.	HVS	Min.	HVS	Basic	Certified
Minimum sample size	500 g	1 kg	500 g	1 kg	500 g	1 kg	500 g	500 g
Total other plant species (max. no. of seeds)	4	1	10	2	10	4	4	10
Other cultivated species (max. no. of seeds)	1†	0	7	1	7	3	1†	7
Non-cereals (max. no. of seeds)	3	1	7	1	7	2	3	7
Wild oats or darnel (max. no. of seeds)	0*	0	0*	0	0*	0	0*	0*
Corn cockle or wild radish or, for HVS only, sterile brome or couch (max. no. of seeds)	1	0**	3	1	3	1	1	3
Ergot (max. no. of pieces)	1	0	3	1	3	1	1	3
Minimum % purity	99	99	98	99	98	99	98	98
Minimum % germination	85	85	85	85	85	85	85	85
Maximum % moisture content	17	17	17	17	17	17	17	17

Notes:
† 2 seeds in 500 g shall not be considered an impurity if a second 500 g from the same sample is free.
* 1 seed in 500 g shall not be considered an impurity if a second 500 g from the same sample is free.
** On basic seed the nil standard applies in repect of wild radish and corn cockle only.

Table 6.4 Oil seeds: seed crop standards. Varietal purity.

Minimum varietal purity (percentage by number)	Basic Seed	Certified Seed or Certified Seed of the First Generation	Certified Seed of the Second or Third Generation
Rape seed*	99.9	99.7	—
Flax and linseed	99.7	98.0	97.5

* For rape of varieties to be used solely for fodder purposes the standards shall be 99.7 per cent for Basic Seed and 99 per cent for Certified Seed.

Table 6.5 Seed standards: oilseeds.

	Oilseed rape Basic	Oilseed rape Certified	Flax	Linseed
Minimum sample size	70 g	70 g	150 g	150 g
Minimum germination % of pure seed	85	85	92	85
Minimum analytical purity, % by weight	98	98	99	99
Maximum content of seeds of other plant species, % by weight	0.3	0.3	—	—
Maximum content by number of seeds of other plant species in the sample:				
total			15	15
wild oat	0	0	0	0
dodder*	0	0	0	0
wild radish	10	10		
dock	2	5		
black grass			4	4
lolium remotum			2	2
maximum number of sclerotia	10	10	—	—

* One seed in a sample shall not be considered an impurity if a second sample is free.

Table 6.6 Peas and beans: field crop standards.

Minimum varietal purity (percentage by number):

Basic	99.7
C1	99.0
C2 & later	98.0

Table 6.7 Peas and beans: seed standards.

	Peas		Beans	
	Basic	Certified	Basic	Certified
Percentages by weight:				
Minimum analytical purity	98	98	98	98
Maximum content of other seeds:				
total	0.3	0.5	0.3	0.5
a single other species	—	0.3	—	0.5
melilotus	—	0.3	—	0.3
Percentages by no. in sample (a)				
Minimum germination	80	80	85	85 (b)
Maximum hard seeds				5
Maximum content of other seeds:				
a single other species	20	—	20	—
wild oats	0	0	0	0
dodder	0	0	0	0
dock	2	5	2	5
melilotus	0	—	0	—

Notes:
(a) Minimum sample size of 1 kg.
(b) Up to the maximum indicated, hard seed shall be regarded as capable of germination.

6.4 Seed labelling

The EC Directives set detailed standards for the labelling of seed. These are intended to ensure that the seed and its source can be properly identified. Official labels are required on all marketed seed (except that sold in small packages). The reference number of the seed lot, and the month and year sealed have to be stamped on the label under the supervision of a licensed seed sampler, at the time of sampling for official examination.

Seed lot reference numbers are constructed as follows:

last digit of harvest year/category code/processor's registered number/ sequential number for the lot.

For example: 9/1H/9876/27 means that the seed is from the 1989 harvest, certified first generation HVS, processed by registered processor number 9876, and is the 27th lot of seed that he has produced from that harvest year.

However, whilst in most cases the harvest year number should be unambiguous, it should be noted that where seed has been bulked, the year shown will be that of the most recently harvested seed. The category codes are shown in Table 6.8.

The sequential numbers are for each lot of seed produced by the processor that harvest year, regardless of the type of category of seed.

Table 6.8 Official seed label colours and category codes.

Colour	Category	Code
Violet	Breeder's	BR
White with violet diagonal	Pre-basic	PB
White	Basic	BS
	Basic HVS	BH
	Basic minimum	BL
Blue	Certified	CS
	Certified HVS	1H
	Certified minimum	1L
	Certified first generation	C1
	Certified first generation HVS	1H
	Certified first generation minimum	1L
Red	Certified Second Generation	C2
	Certified Second Generation HVS	2H
	Certified Second Generation Minimum	2L
	Certified Third Generation	C3
Brown	Commercial	CM
Green	Mixtures	—
	Mixtures HVS	—
Buff	Uncertified Pre-basic (supplier's label, not official)	

The required colours of the labels are shown in Table 6.8 and Table 6.9 lists the categories which apply to different seeds. HVS seed may be marked with a special symbol.

6.5 Seed production aid

Seed production aid is paid on herbage seeds, linseed, flax and hemp seed, and on spelt wheat and rice.

Seed qualifies for aid provided it is produced under a growing contract between a producer and a processor/merchant (or is produced by a grower/processor), it is harvested between 1 July and 31 December, sampled by an official sampler, and certified as basic or certified.

The growing contract, or declaration of growing in the case of grower/processors, has to be received by the Intervention Board by 31 January of the year following the year of harvest. Then the aid claim form, signed by both the merchant and grower, and with accompanying seed certification forms,

Table 6.9 Seed categories in use for different crops.

Wheat, barley, and oats: Breeder's
                                   Pre-basic
                                   Basic (HVS and Minimum)
                                   Certified first generation (HVS and Minimum)
                                   Certified second generation (HVS and Minimum)
                                   Mixtures (HVS and Minimum)

Rye: Breeder's
        Pre-basic
        Basic
        Certified
        Mixtures

Peas and beans: Breeder's
                    Pre-basic
                    Basic
                    Certified first generation
                    Certified second generation

Herbage seeds: Breeder's
                   Pre-basic
                   Basic
                   Certified*
                   Certified HVS and Minimum*
                   Commercial*
                   Mixtures (and HVS mixtures)*
                   (* categories vary according to species)

Oilseed rape: Breeder's
                 Pre-basic
                 Basic
                 Certified

Flax and linseed: Breeder's
                    Pre-basic
                    Basic
                    Certified first generation
                    Certified second generation
                    Certified third generation (flax only)

has to be submitted to the Intervention Board before 30 June. Payment will normally be made to the grower within two months of receipt of the claim, and in any case by 31 July.

Applications for aid are normally made by the contracting merchant, who charges a small fee for the service (e.g. 30p/50 kg).

The rate of aid is fixed, in advance, for two years. The 1988 price fixing established aid rates for the 1990/91 and 1991/2 marketing years. These are shown in Table 6.10.

Table 6.10 Seed aid.

	Aid rate/100 kg		
	1989/90 £[1]	1990/91 and 1991/2 ECU	£[1]
Grasses			
Velvet Bent	44.75	63.8	44.75
Red Top Bent	44.75	63.8	44.75
Creeping Bent	44.75	63.8	44.75
Brown Top Bent	44.75	63.8	44.75
Tall Oat Grass	39.56	56.4	39.56
Cocksfoot	32.12	45.3	31.77
Tall Fescue Fe	34.72	49.5	34.72
Sheep's Fescue	25.25	36.0	25.25
Meadow Fescue	25.25	36.0	25.25
Red Fescue	21.74	31.0	21.74
Italian Ryegrass	12.41	17.7	12.41
Perennial Ryegrass			
High persistence late or medium late	20.62	29.4	20.62
New varieties and others	15.29	21.8	15.29
Low persistence medium late medium-early or early	11.29	16.1	11.29
Hybrid Ryegrass	12.41	17.7	12.41
Timothy	49.24	70.2	49.24
Small Timothy	30.09	42.9	30.09
Wood Meadowgrass	22.94	32.7	22.94
Smooth-Stalked Meadowgrass	22.94	32.4	22.72
Rough-Stalked Meadowgrass	22.94	32.7	22.94
Clovers etc.			
Trefoil	18.73	26.7	18.73
Lucerne (ecotypes)	12.48	17.8	12.48
Lucerne (varieties)	20.62	29.4	20.62
Sainfoin	11.78	16.8	11.78
Alsike Clover	27.07	38.6	27.07
Egyptian Clover	27.00	38.5	27.00
Crimson Clover	27.00	38.5	27.00
Red Clover	29.81	42.5	29.81
White Clover	41.73	59.5	41.73
White Clover (Ladino)	41.73	59.5	41.73
Common Vetch	18.38	25.7	18.03
Hairy Vetch	13.40	19.1	13.40
Oilseeds			
Textile Flax	16.69	23.8	16.69
Linseed	13.19	18.8	13.19
Hemp	12.06	17.2	12.06

Note: [1] Conversion at 1 ECU = £0.701383.

(B) LINSEED AND FLAX

6.6 The subsidies

An outline of the subsidies available is as follows:

(a) Fibre flax aid

An area payment is fixed at the start of the marketing year; half is paid to the farmer and half to the purchaser.

(b) Fibre flax seed aid

This is an area payment, based on average yields, paid either to the purchaser or to the grower. The same crop may qualify for fibre flax aid and the seed aid.

(c) Linseed subsidy

This is an area payment based on average yields. It may be claimed by either the farmer or the processor, but varieties eligible for linseed subsidy are not eligible for the flax aids.

6.7 The rates of aid

(a) Fibre flax aid

Aid for fibre flax is paid on the area under cultivation, at a single rate throughout the Community. The rate is fixed before 1 August, the start of each marketing year.

For the 1988 crop, the total aid (producer and processor) was £239.70. This is subject to a promotional levy totalling £23.97, but the same crop may also qualify for flax seed aid of £169.39 (see below).

(b) Linseed subsidy and fibre flax seed aid

After harvest, average yields are calculated and these are used to determine the linseed subsidy and the fibre flax seed aid. The aids are designed to compensate for the difference between an EC Guide price (fixed before the start of the marketing year) and the average world market price, and the amount paid per hectare is based on the average yield of the area concerned.

Indicative yields are calculated from member states' figures for the quantities of seed produced. A single indicative yield for linseed applies to the whole of the UK and Ireland. Separate indicative yields apply to fibre flax seed aid.

The subsidy rate, in ECU per hectare, is calculated as follows:

(Guide Price – Average world market price) × Indicative yield (te/ha)

This is converted to £s per hectare at the green rate.

For the 1988 crop the rates of aid were:

- linseed: £336.96 per hectare;
- flax seed aid: £169.39 per hectare.

These aid rates were based on a guide price of £374.05 per tonne, an indicative yield of linseed of 1.856 tonnes per hectare (and of flax seed of 0.933 tonnes), and a world price of £192.50 per tonne for 1988/9. The world price was taken from the mean price of Canadian linseed from 5 September 1988 to 3 March 1989.

6.8 Claiming aid

The aid schemes are administered by the Intervention Board, to whom claims must be submitted. These claims are verified by MAFF staff and incorrect claims may lead to a loss of part or all of the subsidy.

It is essential to adhere strictly to the timetable for claims, and ensure that forms are received by the Intervention Board by the due dates or payment will be lost. The timetable for claimining and paying aids that has operated in the past has been:

Linseed:
Year 1:   Sowing declaration submitted by 15 June.
          Harvesting declaration submitted by 15 December.
Year 2:   Aid rate fixed by EC by about March.
          Payment of aid at about April.

Fibre flax and fibre flax seed:
Year 1:   Sowing declaration submitted by 30 June.
          Fibre flax aid rate fixed by EC by 1 August.
          Harvesting declaration submitted by 30 November.
Year 2:   For purchaser to receive second half of flax subsidy, or seed subsidy, purchase contract concluded by 31 July.
          Processor submits purchase contract, or grower submits proof of processing by 31 December. Also, claims for seed aid should be submitted by the same date.
Year 3:   The Intervention Board pay claims by 31 March.

Table 6.11 Varieties eligible for fibre flax aid.

Ariane	Nanda
Astella	Natasja
Belinka	Nynke
Berber	Opaline
Fanny	Regina
Hera	Saskia
Laura	Silva
Lidia	Thalassa
Marina	Viking
Mira	

and any variety currently under study by the
authorities within UK with a view to entering
them in the catalogue of flax varieties
intended mainly for production of fibres.

Aid for fibre flax will only be paid on one of the varieties listed in Table 6.11. Linseed subsidy will only be paid on varieties which are not listed in Table 6.11.

7   Set-aside

7.1 EC set-aside

As part of the 'stabiliser' agreement of 1988, the EC introduced a set-aside scheme, compulsory for member states, optional for individual producers. Participants have to set aside at least 20% of their arable land for a minimum of five years, though with the possibility of leaving the scheme after three years.

Within general guidelines, member states design their own schemes and fix the level of premiums offered. These have to be between 100 and 600 ECU/hectare (£70 to £420/hectare), but may be up to 700 ECU (£490)/ha in special circumstances. Part of the cost is funded by member states, part by the EC.

The EC contribution to the set-aside premiums paid by member states was initially set at 50% for the first 200 ECU, 25% from 200 to 400 ECU, and 15% from 400 to 600 ECU. In September 1989 the Council decided to increase this rate of funding to 60% for the first 300 ECU/ha and 25% for the part of the premiums between 300 and 600 ECU/ha.

The EC's contribution to the cost of set-aside schemes has been increased in the hope that this will stimulate the introduction of more effective national schemes. Only four member states had introduced set-aside schemes in accordance with the timetable set by the EC, and by the summer of 1989 Denmark had still failed to implement any scheme. Portugal is exempt from the scheme, as are a number of regions in other member states where there is a risk of depopulation, or where the natural conditions make it inadvisable.

The effect of set-aside schemes in the 1988/9 marketing year is shown in Table 7.1. The highest uptake was in Germany, which had previously operated a pilot set-aside scheme in Lower Saxony. But rates of uptake in France and Belgium were so low that they did not even amount to 0.1% of arable land. Low uptake was attributed to low premium rates. The Commission estimated that by 1989 only Germany, Spain, Italy, Netherlands and Luxembourg were offering rates of premium which, on average, compensated for the loss of income from set-aside.

The EC set-aside arrangements allow member states to make payment for

Table 7.1 EC set-aside uptake.

	Number of applications	Total area set-aside	Set-aside as % of arable land	Average area per applicant
W. Germany	25,289	169,729	2.4	6.7
Italy	9,301	155,606	1.8	16.7
UK	1,750	54,779	0.9	31.3
Spain	518	34,229	0.3	66.1
France	1,002	15,707	—	15.6
Ireland	77	1,310	0.1	17.0
Netherlands	195	2,621	0.3	13.4
Belgium	32	329	—	10.2
EC total	38,164	434,310	0.9	11.3

Note: No statistics are available for Greece, and set-aside was not operated in Denmark, Luxembourg and Portugal. Italian figures are provisional.

'green fallow' which is extensively grazed. This option is not available in the UK, Luxembourg or the Netherlands. In 1988/9 green fallow accounted for 65% of set-aside land in Ireland, 24% in Italy, 7% in Belgium, 5% in Spain and 1% in Germany; a total of 9% of the EC set-aside land. It has been introduced more recently in France.

7.2 The UK set-aside scheme

The main features of the current UK set-aside scheme are as follows:

(a) Area to be set-aside

Participants must set-aside at least 20% of their land which in 1987/8 was planted with 'relevant arable crops'. The list of 'relevant arable crops' includes all main arable crops, fodder crops and herbage seeds, but not potatoes.

(b) Land which may be set-aside

The actual land which may be set-aside must have been under a 'relevant arable crop', or potatoes, or bare fallow, in 1987/8. It must not have been converted to arable cropping from permanent pasture or another non-arable use between 1 January and 30 June 1988, and must be in arable cropping or temporary grassland at the time of the application.

Table 7.2 UK set-aside scheme compensation payments.

	Land in less favoured areas (£s)	Land elsewhere (£s)
Permanent fallow	180	200
Rotational fallow	160	180
Non-agricultural use	130	150
Woodland (in Woodland Grant Scheme)	180	200
Woodland (in Farm Woodland Scheme)	150 (in disadvantaged areas) 100 (in severely disdavantaged areas)	190

(c) The set-aside options

There are three options for the use of the set-aside land:

- fallow;
- non-agricultural use;
- woodland.

Details of the requirements of these options are given in sections (h), (i) and (j) below.

(d) Compensation payment

The compensation payment is calculated per hectare of land set-aside and is made annually in arrears. The rates of payment are shown in Table 7.2. Further details of the woodland payments are given in section (j) below.

(e) Length of agreement

The set-aside agreement normally lasts for five years from 1 October in the year the application is accepted. It is possible to withdraw from the scheme after three years by notifying the local MAFF Divisional Office before 30 June in the third year of the set-aside agreement. It is not then possible to re-enter the scheme on the same land before the end of the original five-year period.

(f) Penalties

Breaking the rules may lead to loss of future payments, repayment of amounts already received, or criminal prosecution for fraud or deception.

(g) Registration

Because the calculations for the set-aside scheme are based on 1987/8, farmers were encouraged to register the details of their farming in 1987/8 before 30 September 1988. This was to facilitate joining the scheme at a later date. Applicants who have not previously registered need to provide independent evidence of their 1987/8 cropping pattern. This can be in the form of certification by a qualified surveyor or valuer who had seen or had evidence of the cropping pattern in 1987/8. This may be verified by cross checks with satellite images or aerial photographs. An Ordnance Survey map will also have to be submitted with the application.

(h) Fallow

Permanent fallow means that the same parcel of land is set-aside for the full five years.

Rotational fallow allows different parcels of land to be set-aside each year as part of the normal arable rotation.

Participants must follow the detailed rules of the scheme; in addition there are non-mandatory guidelines available.

In outline, the basic requirements of fallow are as follows:

(i)   The land must normally be kept in good agricultural condition.
(ii)   Land must not be left bare. A green cover crop must be sown, or established by allowing the naturally occurring vegetation to regenerate.
(iii)   The cover crop must be cut at least once a year.
(iv)   Cultivation, destroying the cover crop, is allowed: after 31 August for the land in rotational fallow; if it is necessary to control weeds, provided a further cover crop is established as soon after as is practically possible; or if a change of cover crop is required.
(v)   The application of fertilisers is prohibited, with limited exceptions. These include prevention of erosion, or where the land is managed for geese.
(vi)   The application of pesticides is prohibited but authorisation may be granted for the use of some pesticides under certain limited circumstances, including the control of thistles, dock, ragwort, or serious infestations of wild oats, couch, black grass or cleavers.

(vii) The land may be managed for environmental or conservation purposes. Creation of a wildlife habitat may give exemption from the requirement to cut the cover crop, but a written plan must be submitted.

(viii) Existing trees, hedges, water courses, ponds and pools on, or next to, set-aside land must be maintained.

(ix) No new drainage systems may be installed.

(i) Non-agricultural use

Under the non-agricultural use option, set-aside land may be used for activities such as tourist facilities, caravan and camping sites, car parks, football pitches, tennis courts, golf courses, riding schools, livery stables, and game and nature reserves.

The uses which are prohibited are any form of agricultural production, mineral extraction (including open cast coal mining) or any permanent building or structure to be used for industrial processes, sale of goods by retail or wholesale, as a storage or distribution centre, housing or other residential use, including hotels or office use.

The exception to the prohibition on buildings is that they may be erected on set-aside land if they are used in connection with one of the uses eligible for the farm diversification grant scheme, and continue to form part of the diversified agricultural business of the holding. Such permitted uses are: certain types of farm-based industry, farm shops for sale of produce from the holding, provision of farm accommodation, food and drink, provision of educational facilities relating to farming and the countryside and to farm-based industry, provision of livery, and of horses and ponies for hire; or letting of land or buildings for any of these purposes.

(j) Woodlands

The annual payments (listed under (d) above) are payable for five years in the case of set-aside to woodland under the Woodland Grant Scheme, and for between 10 and 40 years (dependent upon type of planting) for set-aside through the Farm Woodland Scheme.

In addition, grants are available under both the Woodland Grant Scheme and the Farm Woodland Scheme. These are payable in three instalments: 70% on completion of planting, and 20% and 10% after 5 and 10 years respectively, subject to satisfactory maintenance. The rates of grants payable by the Forestry Commission for approved planting under the Woodland Scheme are given in Table 7.3. The planting grants payable by the Forestry Commission under the Farm Woodland Scheme are on a maximum of 40 ha

Table 7.3 Woodland Grant Scheme.

Area approved (ha)	Grant (£/ha) for	
	Conifers	Broadleaved trees
0.25 – 0.9	1005	1575
1.0 – 2.9	880	1375
3.0 – 9.9	795	1175
10.0 and over	615	975

per holding and a minimum of 3 ha per holding, but planting on the whole holding is not allowed. Subject to availability, the payments are as shown in Table 7.4.

Applicants wishing to set-aside to woodland need to make application for the Woodland Grant Scheme or the Farm Woodland Scheme, as well as the set-aside scheme.

(k) Other rules

Some of the other points of importance in relation to the set-aside scheme are:
(i) Taxation
Payments are liable to tax as income. Discussions are taking place on liability for VAT. In any case this will not arise before April 1991. In some cases change of use of land may give rise to local authority rate liability.
(ii) Cereals co-responsibility levy
Setting aside at least 30% of relevant arable land allows the participant to be reimbursed each year for cereals co-responsibility levy on the first 20 tonnes.
(iii) Extension of set-aside area
Possibilities exist for increasing the amount of set-aside land during the course of the set-aside agreement.

Table 7.4 Farm Woodland Scheme.

Area approved (ha)	Grant (£/ha) for	
	Conifers	Broadleaved trees
1.0 – 2.9	505	1375
3.0 – 9.9	420	1175
10.0 and over	240	975

(iv) Change of use
During the first three years of the set-aside period, application may be made to change the use of the set-aside land (unless converted to woodland).
(v) Change of ownership
If the land changes ownership, and the new occupier does not choose to continue the set-aside scheme, there may be a liability to repay amounts received.
(vi) Tenants
Special conditions apply to tenants. Landlord consent must be obtained for participation in the woodland or non-agricultural use schemes.

(l) The Countryside Premium

In addition to the set-aside payments, Countryside Premium is available to participants in the set-aside scheme in the following counties in the East of England: Bedfordshire, Cambridgeshire, Essex, Hertfordshire, Norfolk, Suffolk.

The Countryside Premium is a scheme that provides incentives to farmers for positive management of set-aside land, for the benefit of wildlife, the landscape and the local community. To be accepted into the Countryside Premium scheme, land must have been accepted into the MAFF Set-Aside scheme as permanent fallow.

Annual payments are made for five years to farmers who agree to adopt specific land management options (which may be different on different parcels of land). Those who entered the set-aside scheme in 1988 were able to apply for the Countryside Premium payments for the remaining four years of their set-aside agreement.

The premium payments for different land management options are as shown in Table 7.5.

In addition, participants may in the first year of their Countryside Premium agreement apply for capital payments for certain items. Fixed cost grants are, for example: hedge planting £3 per metre; shrub and tree planting

Table 7.5 Countryside Premium.

	Payment per hectare
Wooded margins	£85
Meadowland	£120
Wildlife fallow	£45
Brent geese pasture	£90
Habitat restoration	varies

£450 per 0.25 ha, or 72p per plant; field gates £170; styles £65. The closing date for 1989 applications was 31 October 1989.

(m) Advice

Applications for set-aside are dealt with by MAFF Divisional Offices.

Details of the Countryside Premium are available from: The Countryside Premium Unit, c/o Countryside Commission, Terrington House, 13–15 Hills Road, Cambridge CB2 1NL (Tel: 0223 354462).

7.3 Extensification and conversion

Community legislation agreed in 1988 established the framework for set-aside, extensification and conversion from surplus products. The extensification and conversion schemes have yet to be implemented in the arable sector, but a Commission regulation has been agreed to provide the framework for extensification schemes.

A future arable extensification scheme may be expected to operate within the following broad rules:

(a) To qualify for aid, producers have to reduce their output of one or more of the eligible products. The list of eligible products includes cereals, oilseed rape, peas and beans, vegetables and fruit.

(b) Member states may provide for two methods of reducing output:

- a 'quantitative' method where the reduction of at least 20% in output is calculated for each individual holding and
- a 'production' method which involves the conversion to a farming system which has been demonstrated beforehand to lead under normal circumstances to a reduction in output of at least 20%. This option may allow conversion to organic farming to qualify for extensification aid.

(c) The UK Government will have to determine the rules for the length of the extensification undertaking and the previous reference period which will be used to assess normal average annual output.

(d) The extensification aid will only be paid if the holding has been farmed by the producer for a minimum period and will continue to be farmed by him throughout the period of the undertaking.

(e) The Commission may authorise a member state to determine specific conditions for the granting of aid in areas where production, or production systems, are already extensive.

The maximum aid payment per hectare which is eligible for assistance from the European Community is 180 ECUs, in the case of annual crops.

Under Community regulations as currently drafted, member states do not have an obligation to introduce an extensification scheme until 1 January 1991, although pilot schemes may be operated in the meantime.

8 Europe's Money

8.1 The European Community Budget

(a) EC's own resources and limits on expenditure

For many years, the European Community's financial resources – termed 'own resources' – came from:

(1) VAT collected by member states, up to a specified maximum (currently 1.4%);
(2) customs duties;
(3) agricultural levies (including MCAs and sugar levies).

Whilst gross contributions to EC funding have been applied in a similar fashion to all member states, the net financial burden has in practice not been equally shared. In particular, it has borne heavily on the UK. In part, this is because the UK has traditionally been a substantial importer of foodstuffs, and has hence had to pay over to Brussels a relatively large sum in levy proceeds. But the major factor has been the low level of its receipts from EC farm support, which has in the past accounted for a very high proportion of total Community spending.

The extent of the UK's net contribution was reduced by the rebate mechanism agreed at Fontainebleau in 1984, but this still left it as one of the major financers of the CAP. As a consequence, the UK Government has taken the lead in seeking financial restraint in CAP spending. Its attempts to contain expenditure have had more effect since the Fontainebleau agreement spread the financing burden more widely, and the increasing inadequacy of the Community budget to meet the various calls upon it has been a further stimulus to reform.

As a result, the funding arrangements were revised in 1988. Member state contributions were increased by the addition of a 'fourth resource' related to gross national product (GNP). At the same time, an overall ceiling of 1.2% of GNP was introduced for Community expenditure.

The sources of revenue for the EC's 1989 budget are detailed in Table 8.1.

Table 8.1 1989 EC revenue.

	Billion ECU
VAT own resources	26.219
GNP-based own resources	3.904
Customs duties	9.954
Agricultural levies	1.277
Sugar levies	1.185
Miscellaneous and surplus from previous years	2.299
Total	44.838

However, the 1988 decisions to increase the Community's own resources were accompanied by some important restraints on agricultural spending:

(1) Introduction of the stabiliser mechanisms (described in full elsewhere).
(2) Guidelines limiting farm price support (FEOGA Guarantee) expenditure: 1988 reference base of 27.500 million ECU allowed to grow annually at only 74% of rate of growth of GNP.
(3) FEOGA Guarantee spending forecast to take no more than 56% of EC budget by 1992 compared with 68% in 1987. The estimated annual growth rate to be 1.8% in real terms over 1988–1991 compared with 5% in 1984–1987.
(4) Intervention stock purchases to be written down to sale value and costs brought into the budget immediately (a special budget allocation outside the reference base was agreed to deal with writing down of existing stocks). This means that heavy intervention purchases will have an immediate budgetary impact, whereas in the past this was delayed until the stocks were sold.
(5) Budget management whereby monthly expenditure is monitored by budget chapter compared with the previous three years. If expenditure threatens to exceed forecasts the Commission may take action to remedy the situation, or make proposals to the Council to strengthen the stabilisers. The Council has to act within two months. The budget has separate expenditure chapters for 'cereals and rice', 'sugar', 'oils and fats', 'protein plants', etc. These new arrangements mean that a budgetary crisis may arise in one sector, even if the EC has surplus funds in other sectors.
(6) If decisions by the Council of Agriculture Ministers on the agricultural price fixing are likely to exceed the costs of Commission proposals, final decisions are referred to a special meeting of the Finance and Agriculture Ministers.

No provision has been made for exceptional circumstances, except

Table 8.2 General breakdown of EC budget expenditure for 1989.

	Billion ECU
Agricultural market guarantees	26.741
Monetary reserve	1.000
Agricultural structures	1.552
Fisheries	0.389
Regional policy and transport	4.331
Social policy	3.269
Research, energy and industry	1.461
Co-operation with developing and third countries	1.032
Repayments and reserves	2.912
Administrative expenditure for Commission, Parliament, Council, Court of Justice and Court of Auditors	2.105
Total	44.838

currency fluctuations, for which a reserve of 1000 million ECU is available each year.

(b) Expenditure of the EC

60 per cent of the 1989 EC budget, nearly 27 billion ECU, is allocated to 'agricultural market guarantees' – the support schemes for farm products. A list of the other items is given in Table 8.2. This proportion is scheduled to

Table 8.3 Top ten commodities of Agricultural Market Guarantees Budget (1989).

	Billion ECU
Milk and milk products	4.720
Oils and fats	4.709
Cereals and rice	4.086
Beef and veal	2.589
*Sugar	2.051
Wine	1.466
Sheep and goatmeat	1.454
Fruit and vegetables	1.221
Tobacco	0.975
Protein plants	0.706
Others	2.764
Total	26.741

* Note, sugar expenditure is offset by sugar levy income.

decrease, inevitably leading to a further tightening up of support schemes, and increased emphasis on other areas.

Milk remains as the most expensive commodity in the 1989 budget (see Table 8.3), with the oils and fats chapter of the budget taking a similar amount and cereals in third place. Approaching 40 per cent of the oils and fats budget arises from olive oil support; oilseed rape takes 30 per cent or 1.38

Table 8.4 Oils and fats budget.

	Billion ECU
Olive Oil	1.765
Rapeseed	1.380
Sunflower Seed	0.984
Soya	0.504
Linseed	0.051*
Other	0.025
Total	4.709

* Note: aid for fibre flax is an additional 24 million ECU.

Table 8.5 The Cereals Budget 1989.

	Billion ECU
Export refunds	2.640
Intervention costs	1.358
Production aid for durum wheat, starch, etc.	0.786
Aids to small producers	0.122
Rice	0.086
Co-responsibility levy	–0.906
Total	4.086

Table 8.6 Protein Plants Budget 1989.

	Billion ECU
Peas and beans	0.501
Dried fodder	0.204
Lupins	0.001
Total	0.706

billion ECU (see Table 8.4). The cereal budget is accompanied by a note about the budgetary effects of cereals substitutes: these account for 1.5 billion ECU – or nearly 40 per cent of the cereals support costs – and are more than the cost of intervention (see Table 8.5). The cereals co-responsibility levy is paying for about one-fifth of cereals expenditure.

Support of protein crops is tenth in the spending league – with peas and beans taking 501 million ECU of the total (see Table 8.6).

(c) Procedures for the budget

The Treaty of Rome sets out the rules for agreeing the budget. In outline they are:

(a) The Commission has to present a preliminary draft budget for the new financial year (1 January to 31 December) by the preceding 1 September.
(b) The Council has to establish a draft budget and forward it to the European Parliament by 5 October.
(c) The European Parliament has 45 days in which to propose any amendments, otherwise the budget is deemed to be finally adopted.
(d) If the European Parliament proposes amendments, they may be accepted or rejected by the Council. If they are accepted, the budget is deemed to be finally adopted. If, however, the Council changes the European Parliament's modifications, the new draft is forwarded again to the European Parliament which has 15 days to react, or else the budget is deemed to be adopted. The European Parliament can, by the majority of its members and three-fifths of votes cast, amend or reject the Council's modifications.
(e) The European Parliament can, by a majority of members and two-thirds of the votes cast, reject the draft budget and ask for a new draft.
(f) If at the beginning of the financial year, the budget has not been voted, a sum equivalent to not more than one-twelfth of the budget appropriations for the preceding financial year may be spent each month. The amount may be determined by budget chapter or other budget subdivision, and must not be greater than that implied by the draft budget. The new budget year has frequently started on this 'twelfths' basis, which may cause difficulties in some areas; for instance cereal exports could be limited.

8.2 The ECU, green currencies and MCAs

(a) The ECU

The European Community currency is the European Currency Unit (ECU). The market value of the ECU is determined by the value of the currencies of

Table 8.7 Weighting of national currencies in the ECU.

	Percentage
German mark	30.1
French franc	19.0
Pound sterling	13.0
Italian lire	10.15
Dutch guilder	9.4
Belgian franc	7.6
Spanish peseta	5.3
Danish krone	2.45
Irish pound	1.1
Greek drachma	0.8
Portguese escudo	0.8
Luxembourg franc	0.3

all the individual EC member states weighted (see Table 8.7) according to their importance (termed the 'basket').

As the values of member states currencies fluctuate, so the value of the ECU changes. The exchange rate between the ECU and individual currencies will also vary. This means that, if EC support prices were converted from ECU to national currencies at the market rate, they could vary, perhaps daily. This in turn would create uncertainties for farmers and traders.

(b) Green currencies

Although green exchange rates were first introduced before the major currencies had floating exchange rates (because France needed to ensure that a devaluation of the franc did not lead to higher food prices), they also enable the problems arising from variations in exchange rates to be overcome.

This is because they are fixed rates of exchange, which are used only in the agricultural sector, to convert ECU into national currencies. The rates are usually fixed for the various member states by the Council of Ministers at the CAP price fixing each spring, though changes may occasionally be made at other times.

When new green rates are set, the green exchange rate should move closer to the market rate. However, this posed problems for Germany, because the continued strength of the German mark meant that the green mark had to be constantly revalued, and German farm support prices reduced as a consequence.

In order to prevent this, it was decided in 1984 that the value of the ECU applied to agriculture would be increased in line with the appreciation of the

German mark. This led to the creation of the Green ECU, which is now used as the basis from which agricultural support prices are translated into national currencies.

To calculate the value of the Green ECU, the ECU has to be multiplied by a corrective factor, which at present increases common prices by 13.7%.

(c) The green pound gap: devaluation and revaluation

As green rates of exchange are fixed, and market rates vary, differences arise between the two rates. This is termed the 'green gap'.

Member states may have different green rates for different commodities – for instance, the UK has had separate rates for arable crops, beef, sheep, pigmeat, poultry and dairy products. As a result, there can be a number of different green pound gaps.

The green pound gap may be either positive or negative:

- A negative gap means that UK prices are lower than they would otherwise be, and may be closed either by a devaluation of the green pound – which will increase farm support prices – or by an increase in the market value of sterling.
- A positive gap means that UK prices are higher than they would otherwise be, and may be closed either by a revaluation of the green pound – which will lower farm support prices – or by a fall in the market value of sterling.

Green gaps are expressed in percentage terms. A negative green pound gap of 10% means that farm prices are 10% lower in the UK than they are in other countries with no gap.

A devaluation or revaluation may be expressed in terms of a percentage change in the green currency value, or a change in the percentage points of the green gap or MCA. The effect upon prices may be different according to the way in which this is quoted.

(d) National green rate policies

Because green rates and market rates differ, national support prices based upon a country's green rate will be different from those which would apply if its market rate had been used. Since the UK joined the Community, sterling has for the most part been weak. There has therefore been a tendency for 1 ECU to be worth more when converted into pounds at the market rate than when converted at green rates. This means that support prices have tended to be lower than if market rates had been used. In Germany, which has had a relatively strong currency, the reverse has tended to be true.

Movements in a country's currency, and the way in which its Government

Table 8.8 UK green exchange rates.

	(1) ECU/pound
1 February 1973	2.1644
7 October 1974	2.0053
3 March 1975	1.96178[2]
4 August 1975	1.86369
3 November 1975	1.7556[3]
1 August 1977	1.70463
1 August 1978	1.57678
9 April 1979	1.90625
1 August 1979	1.72039
1 October 1979	1.70148
1 August 1980	1.61641
1 July 1986	1.59491
1 July 1987	1.52405
1 January 1989	1.48133
1 July 1989	1.42575

Notes:
[1] Exchange rates with the ua, until 9 April 1979, when the ECU replaced the ua.
[2] For cereals this rate was only used to calculate levies, refunds etc., not intervention prices.
[3] For common wheat and wheat products the implementation was delayed until 1 July 1976.

reacts by requesting green rate changes, are crucial to determining price levels. Different countries have responded differently.

In the UK, successive governments have made only small changes in the green pound rate, even when the market rate for sterling has fallen considerably. The negative green gap has been as high as 45%. Green pound exchange rates are listed in Table 8.8.

(e) Monetary compensatory amounts

The differences between green rates of exchange and market rates has created the need for monetary compensatory amounts (MCAs). If these were not applied, it is probable that substantial quantities of produce would be moved from one country to another to take advantage of different exchange rates. For instance, so long as the UK has a large negative green pound gap, it would be worthwhile in the absence of MCAs to send cereals to German intervention stores for payment in German marks which could then be converted to sterling at market rates. This would mean that a higher payment was obtained than that available by selling directly into UK intervention stores.

The following simplified example illustrates how the MCA system is designed to neutralise green gap differences:

Assume: an intervention price of 100 ECU/tonne

100 ECU = £50 at green rates.
100 ECU = £60 at market rates.
100 ECU = 150 DM at green and market rates.

Payment for sale of 1 tonne into UK intervention stores = £50.
Payment for sale of 1 tonne into German intervention stores = 150 DM.

150 DM can then be converted to £60 in the money markets, giving a net gain of £10 for selling into the German store. To neutralise this, transfers to Germany would need to be charged an MCA of £10/tonne. Conversely, transfers from Germany to the UK would attract an equivalent MCA payment.

MCAs are calculated weekly. The actual amount is calculated by deducting 1.5% ('the franchise') from the green gaps. For this reason, the MCA percentage is always slightly different from the green gap percentage.

For administrative convenience, adjustments are only made to an MCA if the gap changes by at least 1 percentage point.

In general, MCAs are applied to all commodities subject to intervention and to products derived from such commodities. They apply to trade within the Community and also to trade with third countries.

(f) MCAs and trade

The MCA actually payable or receivable by traders on intra-community trade will be determined by the green gaps of both the exporting country and the importing country.

Although MCAs are designed to correct distortions created by currency fluctuations, the system is imperfect. Because MCAs are adjusted weekly, they reflect the previous week's currency exchange rates which may differ from current rates, perhaps markedly at times. Also, the cereal MCA is calculated on 92.5% of the intervention price at the start of the season, not on market prices, and so may be over-generous or under-generous depending on the level of the market price. The operation of the franchise also adds to the imperfections of the system. For all these reasons, the MCAs may not completely eliminate the differences between green and market rates of exchange, and such currency factors may result in significant trade flows between member states.

In order to assist traders, the Community usually allows MCAs to be fixed in advance for third country trade. This may also influence markets, but advance fixing may be suspended if the Commission takes the view that it could have a major effect upon trade.

(g) Monetary differential amounts on pulses and rape

Trade in oilseed rape and pulses is not subject to MCAs, but the subsidy systems operated for these crops are corrected by monetary differential amounts (MDAs).

The support system for oilseeds and pulses pays to the user of these crops the difference between the community support price and the world price. In a country with a weak currency (and therefore a negative green gap), world prices would appear relatively high compared to a country with a strong currency. The MDA is applied to reduce the amount of subsidy paid to allow for the fact that the market value of the crop is higher in that country because of the weakness of the currency. Thus MDAs remove the possibility of obtaining effectively higher support than is intended by the green rates of exchange. The rate of aid which is payable is determined by the country of production of the crop, even if this is traded between member states.

As with MCAs, the arrangements for MDAs do not fully avoid trade distortions arising from green money differences. This is notably because they have to be fixed on the basis of the previous week's exchange rates, and also because they are not based on market prices. Prefixing of subsidies and MDAs is normally allowed so that traders can operate with greater market certainty, but this arrangement can in itself affect the market.

(h) Phasing out of MCAs by 1992

In 1988 the Council of Ministers agreed to dismantle monetary gaps over a four-year period, with the aim of abolishing MCAs after 1992, when the single market is due to come into effect. This will be achieved by green rate devaluations and revaluations proposed by the Commission. But the question arises as to how the Community will decide to deal with currency fluctuations arising after 1992, especially if a number of countries remain with floating currencies outside the Exchange Rate Mechanism. It also remains to be seen whether green rates of exchange will be abandoned as well.

8.3 The European Monetary System

The European Monetary System (EMS) based on the European Currency Unit (ECU) was introduced in 1979 with the objective of creating greater monetary exchange rate stability within Europe. It incorporates an Exchange Rate Mechanism (ERM) to which all member states belong, except the UK, Greece, and Portugal.

Each country in the ERM has a central rate expressed in terms of ECU.

The members of the ERM undertake to keep their exchange rates within $\pm$ 2.25% of their central rates ($\pm$ 6% in the case of Italy and Spain). Such exchange rate stability will depend in the longer term upon economic policies and in the short term may be influenced by intervention on the currency exchange markets.

If a country's currency value falls out of line with its central rate, the central rates may be changed by a general consensus of the participants – 'EMS realignment'. This alters the value of the ECU and hence affects the size of green gaps and MCAs.

Since 1984 a correcting factor has been applied to prevent a country's central rate revaluation increasing positive MCAs which if subsequently removed by revaluation of its green currency would decrease farm support prices in that country. Now, instead, central rate revaluations have the effect of increasing negative MCAs. Thus, for example, increasing the central rate of the German mark will increase both the value of the green ECU and any negative green pound gap (and MCA). Any German MCA will remain unchanged.

UK membership of the ERM would mean that green rates of exchange and market rates would be kept more in line. Assessment of the balance of advantage to agriculture of membership of the ERM also has to take into account effects on fiscal and monetary policies, including the consequent constraints on the scope for management of the economy, and interest rates.

Index